Rogerio de Almeida Vieira

Functional thin films produced by plasma techniques

Rogerio de Almeida Vieira

Functional thin films produced by plasma techniques

Study of titanium nitride and aluminum nitride functional films deposited on AISI M2 and AISI D6 substrates

Imprint

Any brand names and product names mentioned in this book are subject to trademark, brand or patent protection and are trademarks or registered trademarks of their respective holders. The use of brand names, product names, common names, trade names, product descriptions etc. even without a particular marking in this work is in no way to be construed to mean that such names may be regarded as unrestricted in respect of trademark and brand protection legislation and could thus be used by anyone.

Cover image: www.ingimage.com

This book is a translation from the original published under ISBN 978-613-9-61153-9.

Publisher:
Sciencia Scripts
is a trademark of
Dodo Books Indian Ocean Ltd. and OmniScriptum S.R.L publishing group

120 High Road, East Finchley, London, N2 9ED, United Kingdom
Str. Armeneasca 28/1, office 1, Chisinau MD-2012, Republic of Moldova, Europe
Printed at: see last page
ISBN: 978-620-7-61974-0

Table of contents:

"Imagination is more important than knowledge".
(Albert Einstein, German physicist, 1879-1955)

SUMMARY

The current technological applications of materials are vast, especially metals coated with nitride films. In recent years, a great deal of research has been carried out with the aim of optimizing the adhesion of these films to metal substrates in order to broaden their field of application. Nitride films composed of metal cations, especially titanium, zirconium, chromium and aluminum, have proven to be of great interest in applications where surface hardness, corrosion resistance and high shear strength are required. However, the adhesion of these films has been the main limitation for tribological applications that require high surface mechanical effort. This project proposed the study of film/substrate interface thinning techniques by: i) thermally activated diffusion or interdiffusion and ii) gradual variation of the chemical composition of the film-substrate interface (functional films). The films and interfaces formed between TiN and AlN films and AISI M2 and AISI D6 (VC131) substrates were studied. In the first stage, two types of films were produced: titanium nitride and aluminum nitride. The titanium nitride films were obtained by cathodic arc deposition of TiN. The aluminum nitride films were obtained by PVD (physical vapor phase deposition) using magnetron sputtering. These films were characterized by: X-ray diffraction (XRD), secondary electron and backscattered electron microscopy (SEM), energy dispersive X-ray spectrometry (EDX), atomic force microscopy (AFM), Rutherford backscattering spectroscopy (RBS) and 4-point bending tests. The results showed excellent adhesion of the titanium nitride film and poor adhesion of the aluminum nitride film deposited on M2 and D6 steel substrates. In the second stage, functional titanium nitride and aluminum nitride films were deposited. These functional films were obtained using magnetron sputtering deposition. The main objective of this project was to obtain diluted interfaces, which are regions where the properties of the film and the substrate vary gradually. These interfaces formed regions of absorption of the mechanical stresses generated by the interaction between the deposited functional film and the substrate. All the films were characterized by X-ray diffraction (XRD), scanning electron microscopy by secondary and backscattered electrons (SEM), energy dispersive X-ray spectrometry (EDX), X-ray photoelectron spectroscopy (XPS), atomic force microscopy (AFM), Rutherford backscattering spectroscopy (RBS), and 4-point bending tests. The results showed excellent adhesion of the titanium nitride functional film and poor adhesion of the aluminum nitride functional film deposited on M2 and D6 steel substrates. The M2 high speed steel substrates showed the best results in terms of film growth and adhesion.

CHAPTER 1

INTRODUCTION

The field of Surface Engineering has shown great progress in recent years, mainly due to the incorporation of plasma technology into conventional surface treatments and the combination of various other techniques for depositing films of ceramic, metallic and polymeric materials to modify surfaces. It is now possible to obtain film-modified material surfaces using combinations of techniques that result in excellent chemical and physical homogeneity, adequate adhesion of the film to the substrate and excellent control of the thickness of the modified layer[1-76].

Steels and their alloys with modified surfaces are widely used in a variety of important technological applications. These include cutting tools, machine components, stamping dies, where the surfaces of these materials are subjected to great physical stress and/or chemical erosion[1,2,5,7 9,13,15,17,18,22,24,50,55,77,78]. Therefore, layers with suitable properties that adhere to steel surfaces are desirable for these applications. The aim is to increase the mechanical performance, abrasion resistance and improve the corrosion resistance of these surfaces.

Surface modifications of materials by depositing thin films have been used in various sectors of industry with the aim of improving their properties (hardness, resistance to corrosion and shear, coloring). There are various techniques available which are used according to the performance needs of the end product. In particular, the use of plasmas for surface treatment has evolved significantly in recent years. Plasma is a highly energetic medium made up of charged particles (ions and electrons), and possibly neutral particles, which allows physical, chemical and physicochemical phenomena to occur under conditions of metastability, which are not normally possible at ambient temperatures. In processes involving the deposition of films, the main limiting factor is the adhesion of the film to the substrate, due to the formation of a non-coherent interface with different properties to those of the film and the substrate, which is incapable of absorbing tensions, resulting in weak or even non-existent adhesion.

Recent research has shown that techniques using plasmas are very efficient for reactive deposition, as the plasma (ionized gas) normally has a high degree of ionization, emitting ions with multiple ionization states and high thermal energy. These characteristics allow for better adhesion of the film to the substrate, a high reaction rate, excellent uniformity, a high deposition rate, and the formation of films with the substrate at relatively low temperatures.[4,13,17,27,28,66,70,73,79- 89]

The research and development of films of various metals rearranged in the form of multilayers, with the aim of increasing the mechanical resistance of cutting tools and seeking new properties for coatings, has shown excellent results[37-57,71]. Nitride films, formed from the reaction of nitrogen with most refractory materials, have high hardness and are mainly used on cutting tool surfaces to increase their durability. In addition to these applications, the possibilities of using nitride films to coat mechanical parts where wear due to friction is very high have been investigated. In particular, titanium and aluminum nitrides have been used for this purpose and have proved effective in increasing the useful life of cutting tools. The most promising results come from combining these films with other metals in the form of multilayers in order to obtain better tribological and corrosion protection properties[38,39,52- 68,71].

Due to the constant need to obtain better tribological properties, greater protection against corrosion and, above all, to optimize adhesion, another way that is currently being studied is to obtain functional films, presented by this thesis work, which consists of the gradual change of chemical elements present in the film, from the surface of the substrate to the surface of the desired final film. This gradual region of film component elements, located between the substrate and the desired final film, is also called the designed interface, and its main function is to dampen the intrinsic and residual tensions of the film-substrate system.

In surface modifications by film deposition, substrate surfaces need to be properly characterized in order to understand and control the mechanisms that form interfaces. A characterization of these modified surfaces must take into account all the important parameters for each application. In the case of using the treated materials as cutting tools, a series of surface characteristics must be examined (and appropriate techniques employed): surface morphology and topography by scanning electron microscope (SEM), changes in the surface microstructure due to the formation of metastable compounds, the creation of defects and the state of surface tensions (X-ray diffraction (XRD), X-ray photoelectron spectroscopy (XPS), transmission electron microscopy (TEM) and Raman spectroscopy), chemical analysis of the surface, homogeneity of the modification carried out, cross-sectional concentration profile (Energy Dispersive Spectrometry (EDX), XPS and Secondary Ion Mass Spectrometry (SIMS)) and the determination of elemental composition and depth elemental profile of thin films and materials in general (Rutherford Backscattering Spectrometry (RBS) analysis method)[90-109].

In a master's thesis, carried out at the Associated Laboratory of Sensors and Materials (LAS), of the Center for Special Technologies
(CTE) at the National Institute for Space Research (INPE), the formation of an adherent interface between titanium film and 304 stainless steel was developed and characterized[21] . The results show the formation of a diluted interface resulting from the interdiffusion of Ti in the steel and Fe and Cr in the titanium film. It was concluded that this diluted interface was responsible for the adhesion of the film.

This doctoral thesis presents and discusses techniques for diluting the film-substrate interface by varying the stoichiometry of the nitride formed and by temperature-activated diffusion or interdiffusion.

The TiN and AlN films were deposited on special steel substrates with surfaces modified by adherent thin films of titanium and aluminum, respectively. The aim was to obtain functional TiN and AlN films, the chemical composition of which varies gradually according to the thickness of the film, in order to increase the adhesion of these films to the steel, mainly by diluting their interfaces and consequently relieving the residual mechanical stresses in the films.

The titanium nitride films were deposited by cathodic arc, available from BRASIMET Comércio e Indústria S.A. The aluminium nitride films and the titanium nitride and aluminium nitride functional films were deposited in a magnetron sputtering (RF) plasma device, available from the Physics Department of the Instituto Tecnológico de Aeronáutica (IEFF). These films were characterized using observations of the crystalline phases present by X-ray diffraction, topographical observations by SEM, chemical analysis of the surfaces and interfaces by EDX, roughness measurements by atomic force microscopy (AFM), analysis by backscattered electrons, quantitative analysis of elements and chemical compounds present on the surfaces by XPS, analysis of chemical elements by RBS and adhesion by the 4-point bending method.

CHAPTER 2
SURFACE ENGINEERING

Surface Engineering is an emerging, multidisciplinary field comprising many branches of Engineering, Physics, Chemistry and Materials Science. Its great evolution has been accentuated by advances in coatings and surface treatment, motivated by the realization that the surface, for a large number of applications, is the most important part of components. Many component failures have been associated with phenomena located on the surface, such as cracks, inclusions, grain contours and phase transformations[11,24-32,35,36,38,39,110,111].

2.1. Surface modification of materials

Currently, the modifications that can be incorporated into the surface of materials can basically be classified into two types: 1) with interface formation (film coatings) and 2) without interface formation (surface modifications by thermal, chemical, thermochemical, electrochemical, plasma and ion implantation treatments)[11,21,58] .

The interface can be defined as a region in the structure of materials where changes in physical, chemical and structural properties occur. These differences in properties induce the system to exhibit different chemical, physical and structural behavior between the surface and the rest of the body of the piece (substrate). In the case of films, the interfaces are internal surfaces, formed by the inner surface of the film and the outer surface of the substrate.

In processes involving film deposition (such as: electrochemical, electron beam vaporization/deposition, physical vapor deposition (PVD), chemical vapor deposition (CVD), the main limitation is the adhesion of the film to the substrate[11,21,58] . The formation of an interface with chemical and/or physical properties different from those of the film and the substrate can make it incapable of absorbing mechanical stresses, resulting in weak or even non-existent adhesion. Figure 2.1 schematically shows the formation of a defined interface between the film and the substrate, in which there is a very abrupt variation in the concentration of film atoms in this region. In this case, there will be no adhesion unless there is a very strong chemical interaction between the atoms in the film and the substrate.

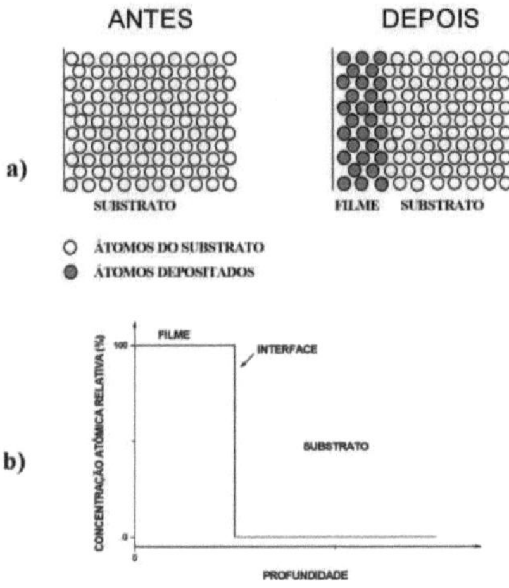

FIGURE 2.1. a) Characteristic distribution of atoms at the film-substrate interface and b) representation of the variation in the concentration of atoms in the film as a function of depth from the modified surface.
SOURCE: [11, 58].

The most commercially used surface modification techniques without the formation of interfaces are: carburizing, nitriding, carbonitriding and boriding. In these processes, the nitrogen, carbon and boron atoms penetrate the surfaces of the materials by diffusion, preferably forming solid solutions, but nitrides, carbides, carbonitrides and borides can also form with the atoms of the substrate's component elements. The characteristic distribution curve

of the chemical elements near the substrate surface is shown in Figure 2.2. Currently, with the advance of plasma technology, the processes mentioned above have resulted in greater homogeneity of the modified surface and efficient control of the thickness of the modified region. The
ion implantation by beam or by immersion in plasmas provide better results, but are not yet economically competitive and are therefore still only used for special applications. In this case, there is no defined interface between the film and the substrate (Figure 2.2). MODIFICATION OF MEV SURFACES FORMATION OF

INTERFACE

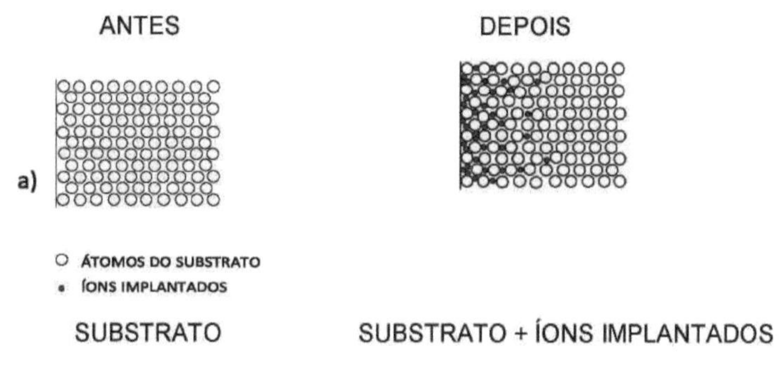

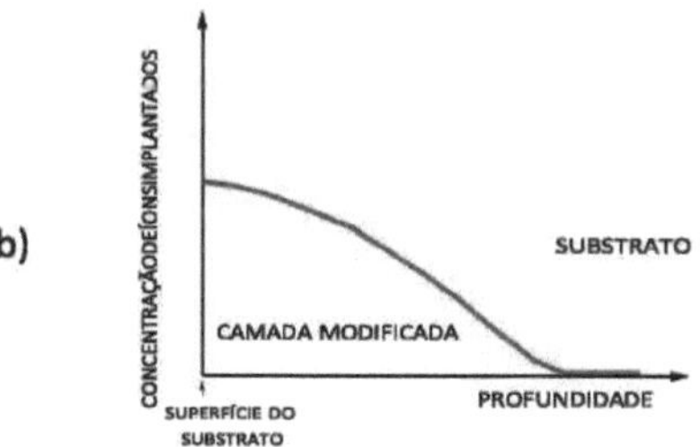

FIGURE2.2 Surface modification of materials with gradual variation of the chemical composition from the surface: a) crystal structure showing the diluted interface and b) characteristic curve of the relative atomic concentration of the dominant element in the film as a function of the penetration depth of the atom from the surface of the film.
of the relative atomic concentration of the dominant element in the film as a function of the penetration depth of the atom from the surface of the film SOURCE: [11, 58].
The combination of atom and/or ion deposition, thermally activated diffusion and/or ion implantation techniques aims to dilute the film-substrate interface, as shown schematically in
Figure 2.3. These techniques, called atom and/or ion mixing, should be applied in situations where it is difficult to achieve dilution of the interface between the deposited film and the substrate. Generally, dilution of this interface is achieved by traditional processes such as thermally activated diffusion or by more sophisticated processes such as ion implantation interspersed with the deposition of very thin films.
DEPOSITION WITH DILUTION OF THE
FILM-SUBSTRATE INTERFACE

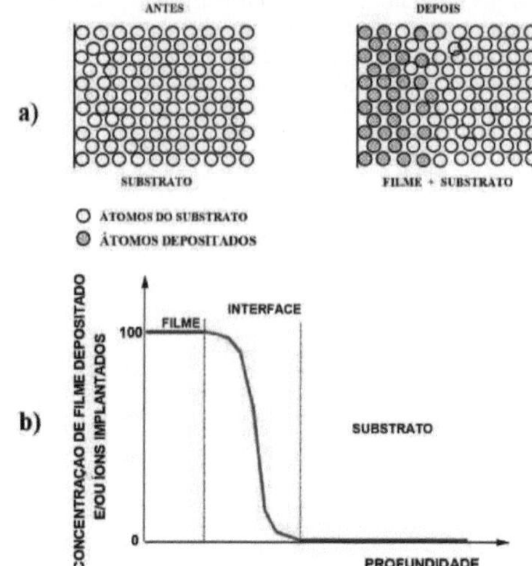

FIGURE 2.3 Combination of film deposition with technique for diluting the film-substrate interface: a) crystal structure showing the diluted interface and b) characteristic curve of the relative atomic concentration for the dominant element in the film versus depth from the film surface source: [11, 58]..

The choice of technique(s) should be based on a number of factors, such as the degree of compatibility of the size and/or charge of the ions involved (for the formation of complete or partial solid solutions), the ability to form chemical compounds between the atoms of the film and the substrate, the degree of crystallographic coherence at the interface, among others[82,112-116] .

2.2. Surface Modification Treatments for Materials with Defined Interfaces

The most commonly used surface treatments for materials with interface formationsão [4,7,8,11,17,21,26-30,35,36,49,79,80,82-84,112-121]:

- Electrochemical treatments; - vapor phase deposition;
- Sputtering deposition and
- Plasma deposition: ion plating, CVD, PVD, CVD/PVD combination, arc deposition.

The main advantages of using surface modification techniques for materials with defined interfaces are:

- Techniques that use low energies for deposition;
- The cost of the equipment is relatively low and
- Most of the staff employed do not need to be highly specialized.

The main disadvantages of surface modification of materials where defined interfaces are formed are:

- Thermodynamics: many metastable compounds are not form at the temperatures and pressures used.
- Physical: certain materials do not diffuse satisfactorily into the substrate structure, causing the film to fail to adhere.
- Environment: some processes produce toxic by-products that are very harmful to the environment.
- Dimensional variations: in many cases where very high tolerances are required, these processes are unsuitable.
- Changes in microstructure;   usually resulting from processes in which undesirable heating occurs (diffusion and/or recrystallization of the film and/or phase changes).

2.3. Treatments for Surface Modifications of Materials without Interface Formation

The most widely used techniques for modifying material surfaces without forming a defined interface between the modified region and the interior of the substrate material are[4,7,8,11,17,21,26- 30,35,36,49,79,80,82-84,112-121]:

- Heat treatments;

- Chemical treatments;
- Thermochemical treatments: nitriding, carburizing, nitro-carburizing, boriding, carbo-boriding;
- Plasma treatments: nitriding, carburizing, nitro-carburizing, boriding, carbo-boriding and
- Ion implantation: ion beam and plasma immersion.

The main advantages of using techniques to modify the surfaces of materials without forming defined interfaces are:
- speed: in some cases the treatment lasts only a few minutes;
- cleaning: results in surfaces free of undesirable residues;
- environment: practically no toxic waste is produced and - dimensional variations: very small.

Techniques involving high energies of ions to be implanted or deposited on the surfaces of materials are not limited by considerations:
- Thermodynamics: allows the formation of stable and/or metastable compounds and
- Physical: allowing the implantation of any atomic species in any material (substrate), as this technique is not limited by the diffusion of the ionic species implanted on the surface of the substrate.

The main disadvantages of material surface modifications that do not promote the formation of defined interfaces are:
- The need for high initial investments and highly specialized personnel;
- In some cases, they depend on the line of sight (ion beam implantation);
- the thickness of the implanted layer tends to be small, in the order of 5 to 500nm and
- Theoretically, any element or chemical compound can be implanted/formed, but in practice, there are difficulties in vaporizing and ionizing certain metals and some chemical compounds at appropriate densities.

Table 2.1 shows the main areas of application, the use and the types of materials involved in each case for the surface modifications shown above.

TABLE 2.1. Applications of materials with modified surfaces.

AREA OF APPLICATION	SPECIFIC USE	MATERIALS INVOLVED
Protective films	Wear protection Corrosion protection Solid lubricant Diffusion barrier Thermal barrier	TiN, TiC, TiCN, TiAIN TiN, BN, CrN, TiB TiN, TiC, MoS TiN, Ti ZrO2(2-15 %p CaO, MgO, Y2O3)
Medicine	Films for biocompatible surgical implants Biocompatible layers with blood Films for intraocular and contact lenses Sterilization of medical instruments	HA (hydroxyapatite), TiN, TO2 Polymers synthesized via plasmas Polymers

		synthesized via H2O2 (hydrogen peroxide) plasmas
Decorative films	Kitchen handles Wristwatch straps, eyeglass frames, cosmetic cases Jewelry and costume jewelry	TiN, TiCN, TO2 TiN, TiCN TiN
Optical films	Scratch resistance Optical filters Reflection control Environmental control Electrical conduction Water-repellent surface Architectural decoration	SiO2 Dielectrics (MgFn AI3O2, SiO2) Al, Ag, Au, Cu, dielectrics SnO2 dielectrics, In2O3, In-Sn-O alloys (ITO) Polymers synthesized via plasmas Metals
Microelectronics	Thin film resistors and capacitors Electrical contacts in integrated circuits Integrated circuits (VLSI, ULSI) Memories for magnetic behavior. Magneto-optical recordings	In2O3, Sn, Ta,Ta2O5 Noble metal silicides (Pd-Si,...) Dry etching $(SF6 + Cl_2)$/ noble gas sputtering Garnets (Y-Fe-O) / rare earth alloys, metal-alloy transalloying
Textile films	Water repellent Wrinkle prevention Antistatic Shrinkage prevention Aesthetic/decorative effects	Polymers synthesized via plasmas Polymers synthesized via plasmas Polymers

		synthesized via plasmas Polymers synthesized via plasmas Metal films
Tribological, cleaning and chemical protection	Surface hardening Surface cleaning Surface control	Nitriding, carburizing Sputteretching Sputteretching
Environmental control	Selective membrane permeabilization Thermal conditioning of buildings	Polymers synthesized via plasmas Metal or dielectric films on glass windows

Due to the current development of plasma technology, many of the modifications listed are already being carried out using processes involving this technology.

2.4. Titanium Nitride (TiN) and Aluminum Nitride (AlN) films

The development of surface modification of materials by nitride films has evolved greatly in recent years, mainly due to the need to obtain films with greater adhesion to metallic substrates, thus increasing their field of application.[22,39,41-47,49,57,59,62].

Among the most widely researched films conventionally used in industry for cutting tools or metal forming, the most important are coatings made from titanium nitride and aluminum nitride films[22,39,4145,47,49,57,59] . Table 2.2 shows some of the physical and mechanical properties of these materials.

Titanium nitride (TiN) is a member of the refractory transition metal nitride family that exhibits properties characteristic of both covalent materials and metallic compounds. The importance of TiN lies in its high hardness and resistance to wear and corrosion, which allow it to be used as cutting tool coatings[39,41,43,50] .

It is the most widely used coating, mainly due to its properties such as hardness, adhesion and resistance to high temperatures. In addition, TiN is a biocompatible material and approved for use by the food industry. Because of its intrinsic biocompatibility, TiN is also a satisfactory material for orthopaedic implants and has been used as a film for hip replacements, dental implants and in surgical tools[60-62] .

Another member of the refractory transition metal nitride family that also exhibits a set of physical properties with excellent characteristics is aluminum nitride (AlN). The most notable and important property exhibited by AlN is its high thermal conductivity. At low temperatures $(= 200°$ C), its thermal conductivity exceeds even that of copper. This high conductivity together with its resistivity and dielectric strength allow it to be used in microelectronic components[59] .

Because it has high thermal conductivity, as well as a number of physical properties, aluminum nitride (AlN) is an excellent candidate to replace alumina (Al2O3) and beryllium (BeO) in the manufacture of high-performance electronic devices[59] .

TABLE 2.2. Physical and mechanical properties of TiN and AlN.

Properties	TiN	AlN

Specific mass (g . cm-3)	5,4	3,28
Tensile strength (MPa)	4930	300-350
Modulus of elasticity (GPa)	250,37	310
Vickers hardness (kg . mm)$^{-2}$	2100	1225
Fracture toughness (MPa . m /)$^{-12}$	5,3	3,35
Coefficient of thermal expansion RT- 1000° C (x10^{-6} K)$^{-1}$	9,4	5,6
Thermal conductivity (W . m^{-1} .K)$^{-1}$	28,84	340
Resistivity in volume (g . cm)	1,3 x 108	1010
Melting point (^{o} C)	2930	>2200
Crystalline form	cubic (cfc)	hexagonal
Color	yellow-bronze (gold)	white

SOURCE:[5 9, 69, 122]

The applications of titanium nitride and aluminium nitride films involve many technological areas, as shown in Table 2.3.

TABLE 2.3. Technological applications of titanium nitride and aluminum nitride films.

AREAS OF APPLICATION	**TYPES OF APPLICATIONS**	
	TiN	AlN

Medical and dental	Coatings for surgical instruments, orthopaedic prostheses, orthopaedic implants	-
	minimizes wear on ball bearings	
	Low coefficient of friction, excellent barrier to atomic and ionic diffusion	-
Chemistry	Chemically stable material	-
	Faster and more thermally efficient electronic devices	
	-	Piezoelectric characteristics
Space	Electronic devices	

SOURCE [22, 39, 41-45, 47, 49, 57, 59, 60-62, 122-125].

Currently, the most widely used methods for producing nitride films are: i) by PVD) and ii) by CVD[88,89] .

2.4.1. Nitride Film Deposition Techniques

There are many techniques for growing nitride films, the most commonly used being magnetron sputtering, plasma assisted chemical deposition (PACVD), ion implantation, dynamic plasma mixing, plasma assisted vapor phase deposition (PAPVD), cathodic arc and electron beam combined with electric arc discharge[62-76,88,89] . In this thesis work, two pieces of deposition equipment were used, one by cathodic arc, whose deposition system is located at Brasimet in Sao Paulo, and the other by magnetron sputtering, allocated to the Physics Department of the Technological Institute of Aeronautics (IEFF-ITA).

2.4.2. Obtaining Nitride Films by Cathodic Arc

Among the various evaporation techniques, cathodic arc evaporation is the most versatile PVD film technology (Figure 2.4). The system is based on evaporating and obtaining metal ions, and the effect produced by a high amperage arc generated on the surface of this metal (cathode)[22,35,36,123-125] . The ions are focused by a magnetic field, accelerated and projected onto the substrate to be deposited through a potential difference between the substrates and the reactor chamber. The kinetic energy of the ions is transformed into heat when they collide with the substrate and maintains the temperature during the deposition phase. The compounds are formed when reactive gases of different natures are introduced into the chamber at low pressures.

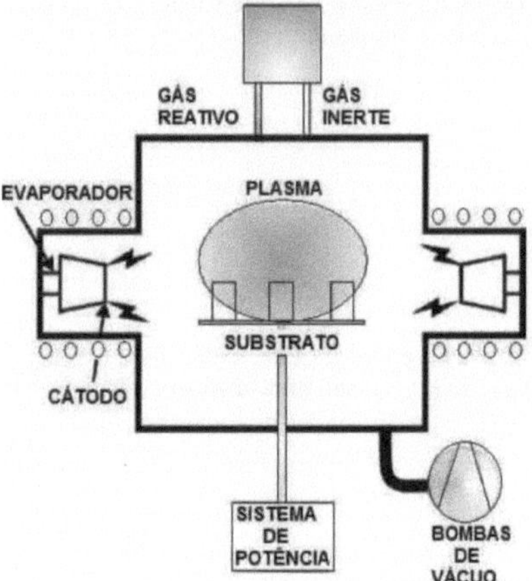

FIGURE 2.4. Diagram of a
cathodic arc evaporation PVD reactor. SOURCE:[123] .

Cathodic arc evaporation generates a high density of ions which makes it possible to obtain large film thicknesses in very short times, as well as making it possible to evaporate different metals simultaneously and alternately introduce different reactive gases.

The process uses arc evaporation to create a highly ionized plasma[124,125] . This allows the production of adhesive films to be applied to substrates at low temperatures. The system operates effectively on a wide range of materials and produces films with high adhesion.

superior performance such as: titanium nitride (TiN), titanium carbonitride (TiCN), chromium nitride (CrN), aluminium-titanium nitride (TiAlN), titanium-aluminium nitride (AlTiN) and zirconium nitride (ZrN).

2.4.3. Obtaining Nitride Films by Magnetron Sputtering

This method basically consists of using plasma generated by sputtering[68,71,120,124,125] . Sputtering is a vacuum process used to deposit very thin films on substrates for a wide variety of commercial and scientific purposes. It is carried out by applying a high voltage across a gas at low pressure (usually argon gas at around 5 mTorr) to create a plasma, which consists of electrons and ion gas in a high energy state[120,124,125] . During sputtering, ions from the energized plasma strike a target, composed of the desired film material, and pull atoms from the target to be launched with enough energy to reach and chemically react with the substrate (Figure 2.5).

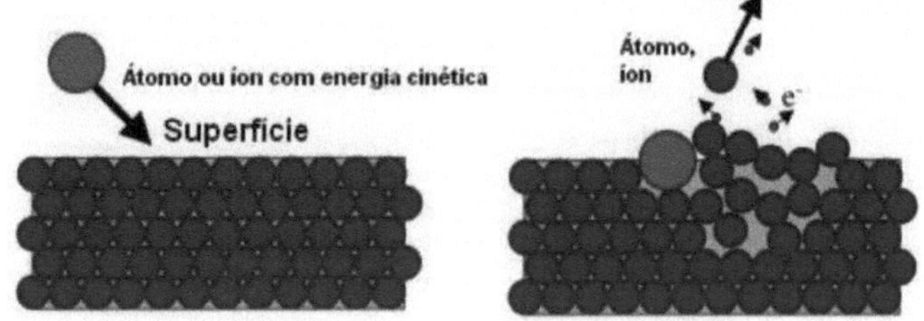

FIGURE 2.5 Principle of the sputtering process. SOURCE:[124] .

2.5.3.1. Magnetron Sputtering

Magnetron sputtering is the most widely used method for depositing thin films in vacuum (Figure 2.6)[62,70,120,124,125] . Although the conventional sputtering method (without magnetron) is still used in some areas of application, magnetron sputtering accounts for 90% of the market for sputtering deposition[120,124,125] . The use of the magnetic field to optimize the sputtering rate leads to the term magnetron sputtering. The deposition rate is generally 10 times higher than the conventional sputtering method. With this system it is possible to produce many types of films with high adhesion and hardness such as: titanium nitride (TiN), zirconium nitride (ZrN), titanium carbonitride (TiCN), titanium aluminium nitride (TiAIN) used in applications such as machine tools and cutting tools.

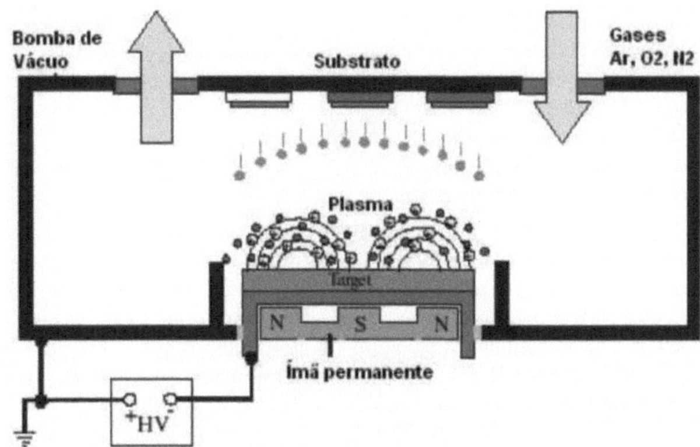

FIGURE 2.6 Schematic drawing of a magnetron sputtering PVD system.
SOURCE:[124] .

2.5 Characterization Techniques for Deposited Films and Interfaces

The evolution of surface characterization techniques has been astonishing over the last two decades. The development and improvement of surface modification processes and new technological applications have created the need for new types of characterization. According to the type of use of each material with a modified surface, an appropriate set of tests is required to characterize it. To characterize surfaces for tribological applications, the characterization techniques that can be used and are available in Brazil are listed[90-109]:

1) Chemical characterization of surfaces
  - EDX (Energy Dispersive Spectrometry) - qualitative and semi-quantitative analysis of chemical elements present;
  - WDS (Wavelength Dispersive Spectrometry) - quantitative analysis of chemical elements present;
  - AES (Auger Electron Spectroscopy) - quantitative analysis of the chemical elements present on surfaces and at depth from this surface (depth profiling);
  - XPS (X ray Photoelectron Spectroscopy) or ESCA (Electron Spectroscopy for Chemical Analysis - quantitative analysis of elements and chemical compounds present on surfaces and at depth from this surface (depth profiling);
  - SIMS (Secondary Ion Mass Spectrometry) - quantitative analysis of elements and chemical compounds present on surfaces and at depth from this surface (depth profiling);
  - LEED (Low Energy Electron Diffraction) - analysis of chemical elements in very thin films;
  - RBS (Rutherford Back-Scattering) - analysis of chemical elements and compounds present;
  - PIXE (Particle Induced X-Ray Emission) - quantitative analysis of the chemical elements present and
  - MET (Transmission Electron Microscopy) - microstructural analysis and analysis of chemical elements and compounds.

2) Analysis of phases and mechanical stress states on surfaces and interfaces
  - Conventional X-ray diffraction - analysis of crystalline phases present in medium and thick films;
  - Low-angle X-ray diffraction - analysis of crystalline phases present in thin films;
  - High-resolution X-ray diffraction - analysis of stresses in films and
  - Raman spectroscopy - analysis of phases, degree of crystallinity and tensions in films.

3) Surface topography analysis
  - SEM (Scanning Electron Microscopy) - observation of defects such as pores, microcracks, roughness, grains and contours and

- AFM (Atomic Force Microscopy) - measurements of surface roughness and porosity in the order of Angstroms.

4) Analysis of mechanical property parameters
- 3 and 4 point bending - film adhesion, mechanical property parameters (Young's modulus, Poisson's modulus, film-substrate shear stress);
- Surface microhardness - hardness and adhesion of films;
- Surface nano-hardness - hardness, adhesion and mechanical property parameters (Young's modulus, Poisson's modulus, film-substrate shear stress);
- Pin-on-disc test - shear test and - Stretch test - film adhesion.

In this work, only the characterization techniques are discussed in more detail: EDX, AFM, XPS, RBS and mechanical testing by three- and four-point bending.

1.1.1. EDX Detection System (Energy Dispersive Spectrometry - EDS)

Every energy dispersive spectrometer has a solid-state detector. Although it can be made of other elements, the detector is almost always composed of a single crystal of lithium-doped silicon, Si (Li)[99,103-105] .

Due to the high degree of perfection of the silicon crystal, its electrons are properly positioned in the crystal lattice. Impurities such as lithium and boron are normally used to make small changes to this structure. This impurity causes free electrons or holes in the crystal lattice. In this case, when an X-ray beam enters the silicon crystal, there is a high probability that it will be absorbed in the interaction with an electron in one of the silicon atoms, thus producing a high-energy electron-photon. The photon-electron in turn dissipates its energy in interactions that stimulate electrons from the valence band to the conduction band, leaving holes in the previously complete valence band. As a result, electron-hole pairs are formed. There is a good statistical correlation between the amount of energy dissipated and the number of electron-hole pairs generated. On average, 3.8 to

3.9 eV of energy are dissipated in the creation of each electron-hole pair[99,103-105].

With this detector, which works in conjunction with the Scanning Electron Microscope (SEM), it is possible to obtain three types of analysis[99,103-105]:

- Spectral curves;
- Image mapping and
- Line mappings.

1.1.2. Atomic Force Microscopy (AFM)

Atomic force microscopy (AFM) is a scanning probe technique in which a tip positioned on a cantilever is scanned in a pattern along the surface[106,107] . Due to forces acting between the surface and the tip, the cantilever is deflected from its equilibrium position (Figure 2.7). It is possible to record the deflection as a function of time and thus form an image of the surface topography with almost atomic resolution.

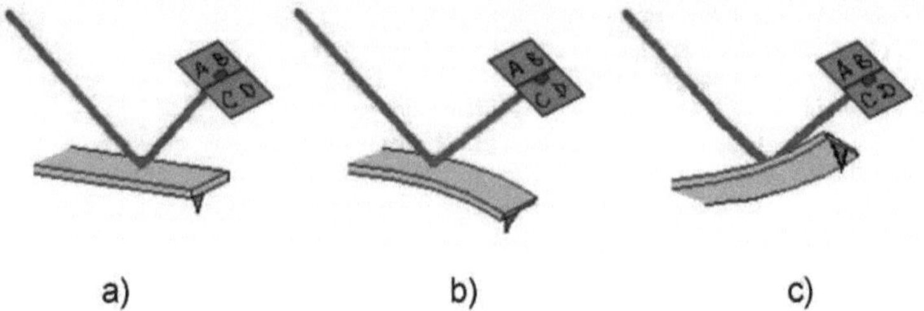

FIGURE 2.7 Cantilever deflections and scanner deformations as a function of the tip-surface interaction force: a) reference force, b) interaction force greater than the reference and c) interaction force less than the reference.

Figure 2.8 shows the basic components of an AFM. The system for detecting the intensity of the tip-surface interaction force is simple, stable and low-noise. This system basically consists of a cantilever, a laser and a photodetector. The laser beam that hits the mirrored upper surface of the cantilever is reflected back to the photodetector, which transfers the information to a computer.

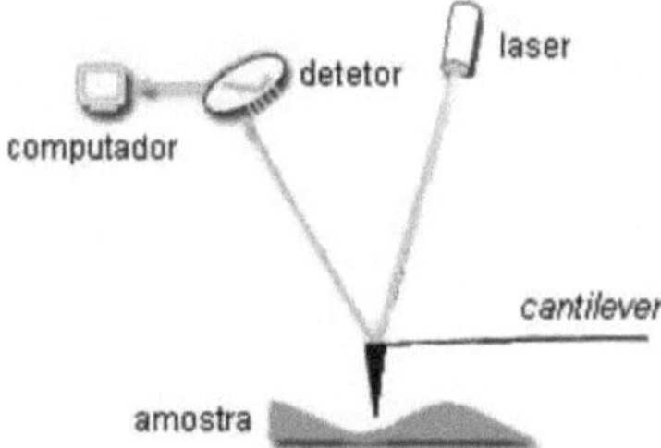

FIGURE 2.8 Basic components of AFM equipment.

The photodetector is divided into four quadrants (Figure 2.7), labeled A, B, C and D. It is possible to monitor the intensity of the laser beam hitting the photodetector at the top (A+B), bottom (C+D), left (A+C) and right (B+D). As the sample is swept under the tip, the intensity of the tip-surface interaction force varies according to the morphology of the sample surface. The deflection of the cantilever and the region of the photodetector hit by the laser beam depend on the intensity of the tip-surface interaction force. The difference between the intensities of the signals hitting the top and bottom of the photodetector is used by the feedback system to define the vertical displacement a

that the sample will be subjected to, in order to keep the intensity of the tip-surface interaction force constant with the reference value.

The photodetectors used in AFM are capable of measuring laser beam detachments as small as 1 nm. In this way, the optical arrangement of the detection system makes it possible to obtain images with lateral resolution in the order of nanometers and vertical resolution of better than 0.1 nm. In atomic force microscopy, the image of the sample surface is formed according to the deformations of the scanner in the x, y and z directions. Each value of the (x,y) pair defines a pixel of the AFM image, which will be associated with a shade of gray defined by the value of z.

Each complete scan of the sample in the x-direction (fast scan) corresponds to a line in the AFM image. Once a scan has been completed in the x-direction, the scanner moves the sample in the y-direction (slow scan) and a new line of the image is Figure 2.9 shows an AFM image of the surface of a TiN film indicating the fast and slow scanning directions.

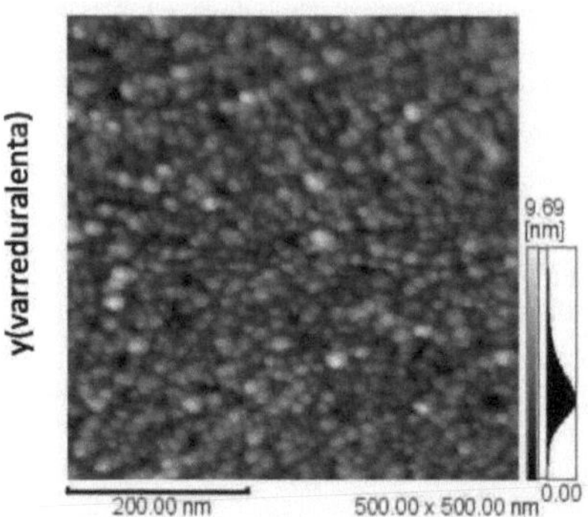

FIGURE 2/9 AFM image of the surface of a TiN film deposited on Si (001).

Several types of tip-surface interaction forces act during the acquisition of an AFM image. The main ones are:

- Van der Waals forces ;
- electrostatic forces   ;
- surface tension forces   and

- Coulombian forces   .

The relative magnitudes of these forces depend on the nature of the sample and the tip, the tip-surface distance, the environment in which the image is being acquired and the operating mode of the AFM. In general, Van der Waals forces are dominant. Figure 2.10 shows the behavior of the Van der Waals force as a function of the tip-surface distance.

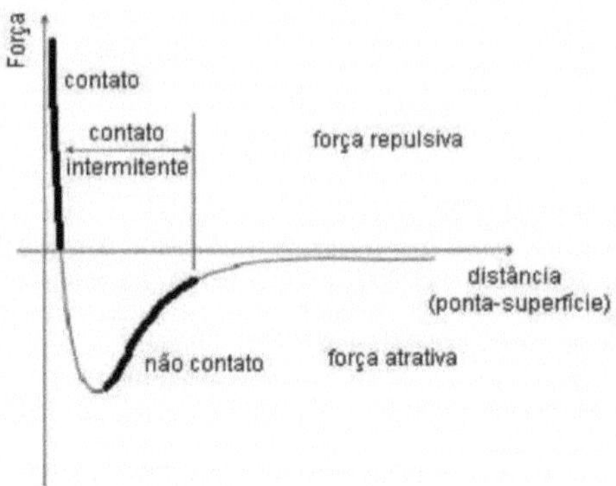

FIGURE 2.10 Van der Waals force between the sample surface and the tip as a function of their relative distance. SOURCE: [106, 107].

When the tip-surface distance is large, the interaction force is practically zero. As the tip and sample get closer, long-range forces start to act and the interaction force becomes attractive (force < 0). Attractive forces result when the tip-surface distance is approximately in the range of 1nm to 10nm. If the tip-surface distance is decreased further, the interaction force becomes repulsive (force >0   ).   The repulsive force results from interactions between orbitals
of the atoms on the surface of the tip and the sample and increases rapidly as the tip-surface distance tends to zero. Repulsive forces act when the tip-surface distance is of the order of a few Angstroms.

Depending on the surface characteristics of the sample and the property you are interested in analyzing, you can operate the AFM in three different modes:
- contact mode (region of repulsive forces);
- intermittent contact mode (region of attractive and repulsive forces) and
- non-contact mode (region of attractive forces).

The working regions of the contact and non-contact modes are illustrated in Figure 2.10.

2.5.3. X-ray Photoelectron Spectroscopy (XPS)

One of the techniques used to study and measure the chemical composition of the surface layer structure is ESCA (electron spectroscopy by chemical analysis), also called XPS (X-ray photoelectron spectroscopy)[108,109].

This technique is mainly used in the following fields of application: corrosion, catalysis, materials, semiconductors, polymers and, fundamentally, materials research.

2.5.3.1. Photoelectric effect

The basic principle of the XPS technique is the photoelectric effect, which can be explained using the energy level diagram shown in Figure 2.11. The bottom three lines, El, E'l and E"1, represent the energies of the electrons in the innermost K and L layers of an atom. The top three lines, Ev, E'v and E"v, represent some of the energy levels of other valence electrons or of a layer.

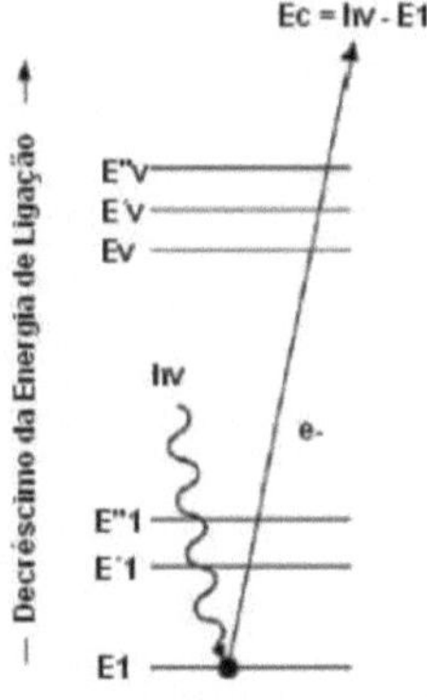

FIGURA 2.11 Schematic representation of the XPS process[126].

In photoelectron spectroscopy, a source of electromagnetic radiation is used to eject electrons from the sample. Two types of conventional photon sources can be used: helium discharge lamp, which produces ultraviolet radiation (hv = 21.2 and 40.8 eV, for He I and He II, respectively), giving rise to ultraviolet photoelectron spectroscopy (UPS), and soft X-rays (hv = 1486.6 and 1253.6 eV for the Ka lines of Al and Mg, respectively), used in X-ray photoelectron spectroscopy (XPS)[108,109]. For surface analysis, XPS is much more important than UPS, as the latter is more specific to electrons from

valencia. The rapid development in instrumentation, interpretation of results and applications has made XPS the most powerful surface spectroscopic technique and it is now used to analyze various types of samples (metals, polymers, ceramics, semiconductor composites and biological samples; in the form of sheets, fibers, powders, particles or films) [108,109].

In practice, it extends from 10 eV, close to the binding energy (13.6 eV) of the electron in a hydrogen atom, to energies around 0.1 MeV. At these energies, photons can penetrate the solid and interact with the electrons inside the surface. Low-energy photons are used to establish the visible spectrum, associated with the outermost weakly bound electrons. These outermost electrons are involved in chemical bonding and are not associated with any specific atom and, consequently, are not useful for identifying the element.

The energy carried by an X-ray photon (hv) is absorbed by the target atom, leading to the origin of the excited state, which is relaxed by the emission of a photoelectron (atom ionization) from the atom's innermost electronic layers[108,109].

The kinetic energy Ec of the photoelectron leaving the target atom depends on the energy of the incident photon, hv, and is expressed by Einstein's photoelectric law,

$$Ec = hv - E_L - \varnothing \qquad , \qquad (2.1)$$

where EL is the binding energy of the photoelectron with respect to the Fermi level and $\varnothing$ is the work-function of the spectrometer, which is a factor that corrects for the electrostatic environment in which the electron is formed and measured[108,109]. In XPS, the intensity of N(E) photoelectrons is measured as a function of their kinetic energies (E_c). However, XPS spectra are usually presented as graphs, in which N(E) is a function of EL.

A technique becomes surface sensitive if the radiation to be detected does not travel more than a few atomic layers (0.5 to 3.0 nm) through the solids. Electrons with kinetic energy between 10 and 1500 eV are ideal for studying surfaces, as their mean free paths in solids are of the same order of energy[108,109]. In XPS, photoelectrons have kinetic energies in the range of 100 to 1400 eV, and when generated close to the surface make this technique very suitable for studying the surface of solids[108,109].

The elements present on the surface of the sample are characterized directly by determining the binding energies of the photoelectric peaks. This is because the energy levels of the photoemission process are quantized. Therefore, photoelectrons have a kinetic energy distribution of discrete peaks relative to the electronic layers of the photo-ionized atom. Figure 2.12 shows an exploratory XPS spectrum.

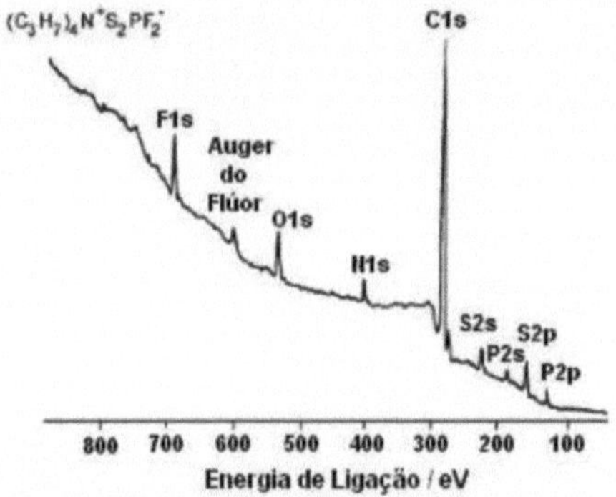

FIGURE 2.12 Exploratory XPS spectrum of a sample
containing various components supported on alumina[127] .

This spectrum shows peaks from a sample containing various components supported on alumina, using magnesium anode radiation[127] . It can be seen that interference from overlapping peaks is insignificant, as the peaks tend to be well separated, even in the case of adjacent elements in the Periodic Table, such as carbon, nitrogen, oxygen and aluminum. It can be seen that the binding energies for the 1s electrons increase with atomic number, as there is an increase in the positive charge of the nucleus (C1s < N1s < O1s), and that more than one peak for a given element can be observed, as is the case with sulphur, which has peaks for the electrons in the 2s and 2p levels.

In practice, the XPS data is   used as if it were characteristic of the atoms as they were before the photoelectric effect took place. However, the photoemission data

represent a state in which the electron has just left$_{átomo}$ [108,109].

2.5.4. Materials Analysis by Ion Beam

The RBS (Rutherford Backscattering Spectrometry) analysis method, together with the PIXE (Proton Induced X-Ray Emission) analysis method, are part of a larger set of nuclear-spectroscopic methods, generically known as Ion Beam Analysis (IBA) methods. These methods have in common the use of monoenergetic ion beams (H^+ , He^+ , He^{2+} , etc.), with energies in the order of a few MeV and tens of nA of current, and are used to determine the elemental composition and in-depth elemental profile of thin films and materials in general. The ion energy of a few MeV/u.m.a. limits the depth analyzed to a few pm. For this reason, these techniques are also known as thin film characterization. When combined, the RBS and PIXE methods make it possible to identify and quantify all the elements in the Periodic Table, except H and He, with detection limits ranging from fractions of a percent to ppm in thick samples and monolayer fraction (~ 1015 cm^{-2}) in thin films.

2.5.4.1.   Backscatter Spectroscopy
Rutherford Backscattering (RBS)

In Rutherford backscattering, monoenergetic particles from an ion beam collide with atoms in a sample, are backscattered and detected by a detector that measures the energy (Figure 2.13). The energy of these particles will depend on the depth of penetration into the sample until collision and backscattering occur and on the mass of the atoms with which they collided[101,102] . In the collision, energy is transferred from the incident particle to the stationary atom. The rate at which the energy of the scattered particle is reduced depends on the ratio of the masses of the incident particle and the target atom and makes it possible to determine the identity of the target atom. Once the target atom has been identified, its specific mass in atoms.cm^{-2} can be determined from the probability of collision between the incident particles and the target atoms, by measuring the total number of particles detected, pd, for a certain number pi, of incident particles. The connection between pd and pi is given by the scattering shock section. Finally, the distance to the surface of the place where the collision occurred can be inferred from the energy loss of the particle on its journey inside the sample. When an ion moves through matter, it loses energy through numerous collisions with the electrons in the material. Due to the small size of the atomic nucleus, the probability of nuclear scattering is very small compared to that of interaction with electrons, and can therefore be disregarded in a first approximation. Since the loss of energy is directly proportional to the length of the path taken by the particle inside the sample, it is possible to establish a depth scale and associate the energy of the detected particle with the location where the collision occurred.

The great success of RBS analyses with H^+ and He^+ beams with energies of around 2.0 MeV is due, in particular, to the possibility of theoretically modeling the experimental spectra with excellent accuracy from fundamental principles, just by assuming classical scattering in a central force field.

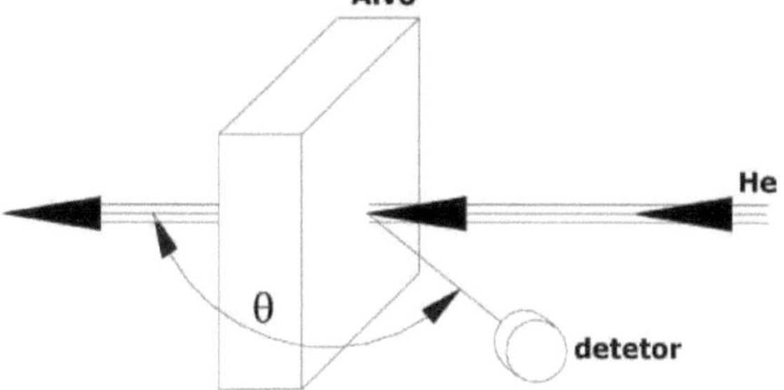

FIGURE 2.13 Representation of the
Rutherford backscattering technique
based on the bombardment of a surface by
alpha particles.

Particles scattered inside a sample lose energy both on the way in and on the way out, towards the detector. In the diagram in Figure 2.14, particles from a beam with energy E0 are incident with angle 01, penetrate to depth Ax and emerge with energy E2 and angle 02 (note that 02 is the supplement to angle 0 defined in Figure 2.14). Angles 01 and 02 are always defined as positive, regardless of which side they are on in relation to the normal axis of the sample.

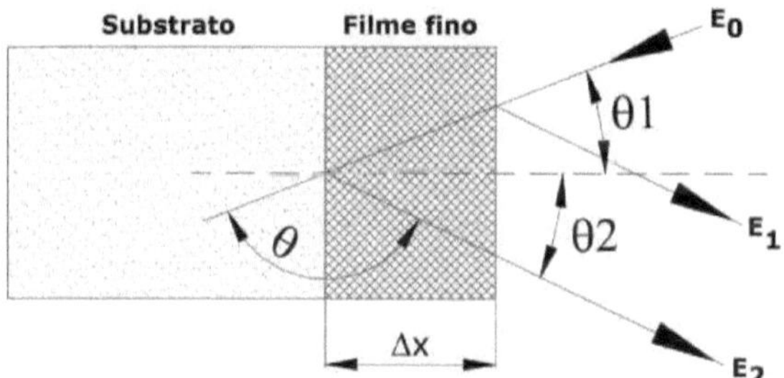

FIGURE 2.14 Schematic representation of
particle scattering.

2.5.4.2.   Interpreting and Processing an RBS Spectrum

An RBS spectrum is a graph of the intensity (count rate) as a function of the energy of the particles detected. A spectrum is actually a histogram, in which the energy axis (abscissa) is divided into 512 or 1024 channels. Each channel corresponds to a small energy interval, of the order of 5 keV/channel. Figure 2.15 shows the spectrum of an RBS analysis of a thin film of a given metal with thickness t, deposited on a silicon substrate. The HM value represents the height of the metal (film) plateau, H_{Si} is the height of the silicon (substrate) plateau, AEM represents the amount of energy lost by the alpha particles backscattered by metal atoms at the metal/substrate interface, E1 corresponds to the energy of the particles scattered on the surface and E2 to the energy of the particles scattered after traveling a distance Ax inside the metal. According to the diagram, the ion beam that hits the sample with energy E0 is scattered on the surface with energy E1 = KME0. When it reaches the interface with the substrate, the energy of the beam is E1 = E0 - AEM. In the first approximation, AEM can be calculated, but in order to determine it, it is necessary to know the film's stoichiometry, which is usually the problem's unknown. In this case, iterative methods and computer programs are used to simulate RBS spectra for a given sample composition/structure[101,102] .

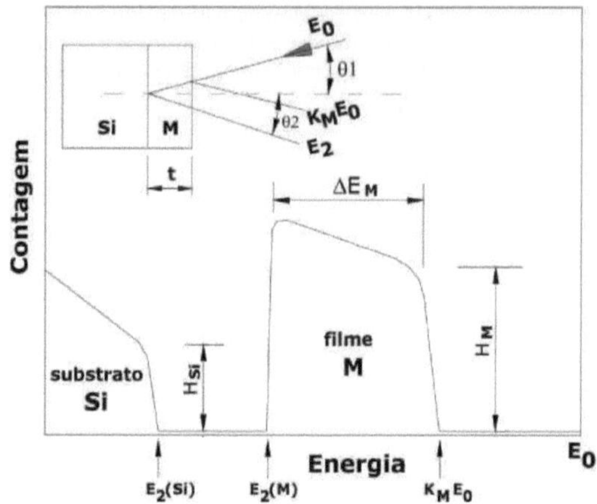

FIGURE 2.15. RBS spectrum of a thin metal film with
thickness t on a silicon substrate.

At the interface, the beam particles are scattered with energy KSiE1 if they collide with silicon atoms (substrate) and KME1 if they collide with metal atoms (film). The scattered particles, each with their own energy, are once again stopped in their tracks.

exit path through the film towards the detector, respectively emerging with energies:

$$E_2(Si) = K_{Si}E_1 - [\eta].t \quad e \quad E_2(M) = K_M E_1 - [\eta].t , \qquad (2.2)$$

where n is the energy loss of a particle with high velocity in a material medium under the conditions and energies of interest in RBS analyses[101,102] .

2.5.5. Three and Four Point Bending Method

Another well-known method, which can also be used to characterize surfaces, is the three- and four-point bending test[128-131] .

This method consists of a long specimen with a circular or rectangular cross-section. The specimen is supported at two points separated by a distance L and is loaded in the center, in the case of three-point bending, and near the ends, in the case of four-point bending (Figures 2.16 and 2.17), respectively. The model for this type of test is given by :

$$\sigma_{máx.} = \frac{M \cdot c}{I} , \qquad (2.3)$$

in which:

M = maximum bending moment;

c = distance from the center of the specimen to one of the ends and I = moment of inertia of the cross-section of the specimen.

In the three-point bending test, the maximum tensile stress occurs on the bottom surface of the specimen and in the same direction as the applied force (Figure 2.16). This results in a concentration of stresses in this direction and rupture tends to occur at the point of maximum stress [128,130,131].

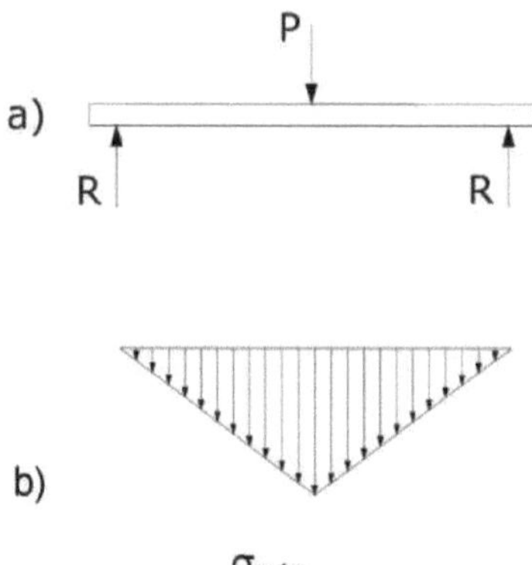

FIGURE 2.16 Schematic drawing of the 3-point bending test on a film substrate: a) position where the load P is applied and the reaction of the two supports and b) distribution of the stress along of the specimen[128,131] .

In the four-point bending test, the maximum tensile stress also occurs on the bottom surface, but over the entire surface between the applied forces, as shown in the schematic drawing in Figure 2.17. This test offers more uniform values, due to the maximum stress being distributed over a larger region of the specimen. prova[129,131].

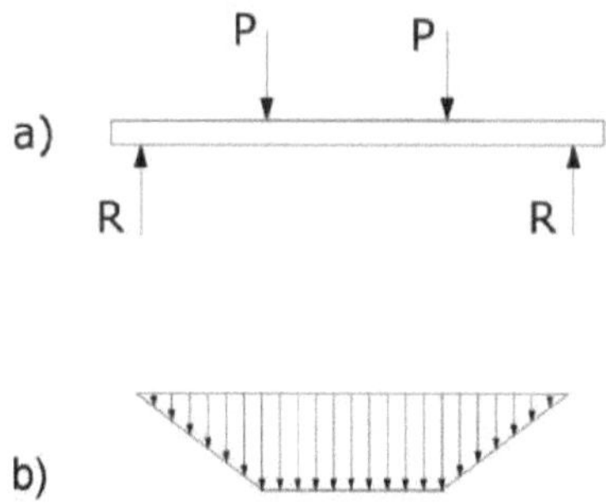

FIGURE 2.17 Schematic drawing of the 4-point bending test for a substrate with film: a) position where the loads P and the two supports are applied and b) stress distribution along the specimen. The maximum stress occurs in the region between the applied loads P[. SOURCE: [128, 131].
This type of bending test makes it possible to obtain fairly reproducible values for the maximum deflection before rupture,máx ., and the tensile strength[128,130 , 131]

CHAPTER 3
MATERIALS AND EXPERIMENTAL PROCEDURE

This work proposes techniques for diluting the film-substrate interface: a) diffusion or interdiffusion caused by temperature; b) functional films, whose chemical composition varies gradually depending on the thickness of the film. The films and interfaces formed between the apo substrate and TiN and AlN films are studied. Figure 3.1 shows, schematically, the

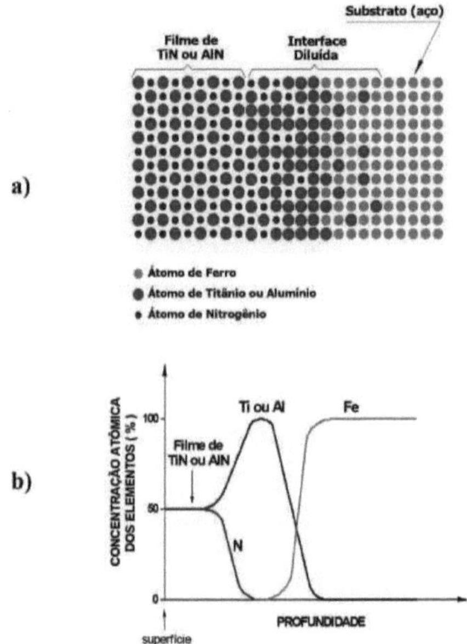

FIGURE 3.1 Schematic proposal for obtaining functional TiN and AlN films on apo substrates using interface dilutions the dilute interface formation project proposed in this thesis project. In order to obtain functional films, deposition should be carried out by varying the metal-N cation ratio in the film, forming stoichiometric and non-stoichiometric compounds resulting in dilute interfaces[58,132] .

The main aim of this thesis is to create dilute interfaces, which are regions where the properties of the film and the substrate vary gradually. This aims to create regions where the tensions generated by the deposited film can be absorbed.

3.1. Materials used

The materials selected for use as substrates in this work were:

1.   M2 high speed steel (AISI M2), with a characteristic chemical composition of: 0.87 % C, 4.20 % Cr, 5.00 % Mo, 6.40 % W and 1.90 % V. This material is generally used in the manufacture of parts that must have high abrasive wear resistance, high compressive strength, medium toughness, good dimensional stability when hardened and excellent hardenability. The main applications for this type of steel are in machining tools, drills and tapers, punches and taps for stamping and cutting, molds for compacting powders, spindles and mechanical components in plastic injection molding, cold forging tools, chopping knives and circular knives[133] .

2.   D6 tool steel (AISI D6), with a characteristic chemical composition of: 2.05 % C, 0.30 % Si, 0.80 % Mn, 12.50 % Cr and 1.30 % W. In general, this material is used in the manufacture of parts that must have excellent abrasive wear resistance, high compressive strength, high surface hardness after heat treatment, good hardenability, good dimensional stability and good temper resistance. The main applications of this type of steel are in stamping tools, cutting tools, drawing tools for material up to 3 mm, circular sheet metal knives, deep drawing tools, bend forming tools, tube and profile forming rollers, metal, ceramic and plastic abrasive powder compaction molds, gauges and in guides and structural components subject to heavy wear[133] .

The M2 quick dies (AISI M2) and D6 tool dies (AISI D6) used in this work were purchased from Uddeholm Apos Especiais Ltda.
The materials used for the deposition, titanium (98%) and aluminum (98%), were purchased by Kurt J. Lesker Company, PA, USA.
The gases used were argon (99.999 %) and molecular nitrogen (99.999 %), both purchased from Air Liquid.

3.2. Experimental Procedure

In this work, an experimental procedure was used which played a fundamental role in the execution and analysis of the deposited films and their characteristics.
Figure 3.2 shows the flowchart of the summary experimental procedure used, followed by the detailed steps.

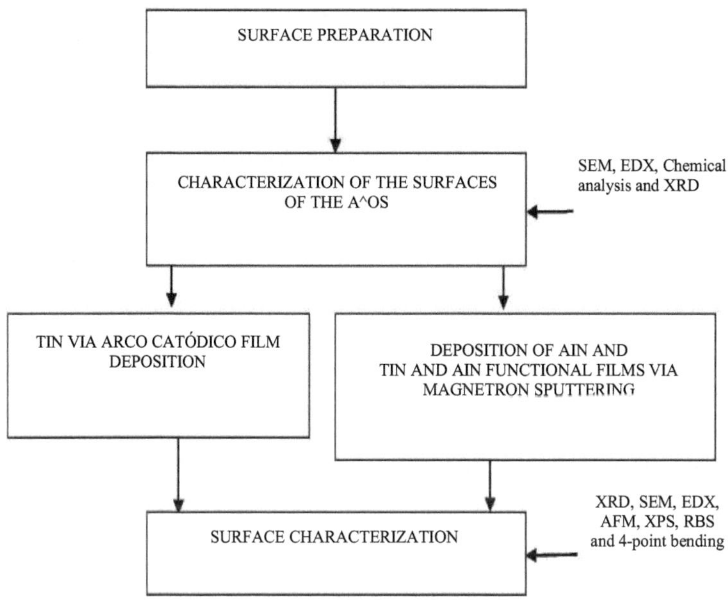

FIGURE 3.2 Flowchart summarizing the experimental procedure used in this thesis.

3.2.1. Surface preparation of the substrates after M2 and D6

The samples of post-fastener M2 (AISI M2) and post-tool D6 (AISI D6) were prepared in the dimensions 20x20 mm by 1 mm thick.
Sanding and polishing of only one of the sample surfaces was carried out on an AROTEC model APL-4D automatic polishing machine.
For the surface sanding procedure, wet sandpaper was used in the following grit order: no. 180, 220, 320, 400, 600 and 1200 (sandpaper grit classification according to ABNT MB481 - Abrasive grits applied to sandpaper - Tests). At each stage of this procedure, the surfaces were sanded until the remaining scratches from the previous sanding had disappeared.
Sandpaper with SiC abrasives (manufactured by 3M) was used.
The samples with one of their surfaces sanded were immersed in detergent and subjected to ultrasonic cleaning for 10 minutes. Then the polishing stage began using cloths (DBM model - supplied by AROTEC) and alumina abrasive powders with particle sizes of 9, 5 and 1 pm. The entire process was carried out wet (distilled and deionized water). Abrasive powders were used in the following order: 9, 5, 1 and 0.5 pm, the latter being

responsible for the final polishing quality of the sample surfaces (roughness). This roughness was measured using AFM and the values were approximately 0.0046 ±0.001 pm.

MICROVIT brand alumina abrasive powders were used to polish the cloths; Type: WCA US.PAT. 3,121,623, from Micro Abrasives Corporation Westfield.

3.2.2. Characterization of Substrate Surfaces

Before depositing the TiN and AlN films, tests were carried out to characterize the surfaces prepared according to the previous item. The aim of this procedure was to study the surface of the steels before the films were deposited.

To characterize these surfaces, the following analyses were carried out: i) topography and microstructure by SEM observations, ii) qualitative and semi-quantitative chemistry using spectral curves obtained by EDX, iii) atomic composition by chemical analysis and iv) the phases present by X-ray diffraction.

3.2.2.1. Topographical and microstructural analysis by SEM

Topographic analyses were carried out on samples of M2 and D6, prepared according to Item 3.2.1, at 1000X and 5000X magnification.

This work used a JEOL JSM-5310 Scanning Electron Microscope (SEM) with secondary detectors for acquiring images and an EDX detector for acquiring spectral curves and mapping[103] .

3.2.2.2. Chemical Analysis of Surfaces by EDX

A substrate of apo M2 and another of apo D6 were used to analyze the chemical elements present, using the spectral curve obtained by EDX.

The detector used in this work was a lithium-doped silicon detector that is part of the JEOL JSM- 5310 Scanning Electron Microscope[103] .

3.2.2.3. Chemical Analysis of Atomic Compositions

Chemical analyses of M2 and VC131 were carried out by direct combustion to determine C; by gravimetry to determine C

determination of Si and W; and by atomic absorption spectrometry of the other elements.

The C and S determiner used was a LECO model CS 200. A VARIAN SpectrAA-20 Plus atomic absorption spectrometer was also used.

3.2.2.4. Analysis of crystalline phases by X-ray diffraction

All the surfaces of the prepared substrates were analyzed by X-ray diffraction to check the surface conditions before the films were deposited.

The X-ray diffractometer used was a Philips model PW3710, with a Ni-LFF filter, Cu anode with À=1.54056 Â (Ka radiation). The samples were analyzed using the following parameters: -   40kV voltage;

-      20mA current; and
-      scan on 29 from 30º to 100º .

3.2.3. Deposition of Titanium Nitride and Aluminum Nitride Films

TiN and AlN films and functional TiN and AlN films were deposited on AISI M2 and AISI D6 steel substrates.

3.2.3.1. Deposition of Titanium Nitride Films

The titanium nitride films were deposited in a 304 stainless steel vacuum chamber, 900 mm high by 650 mm in diameter, with a gas inlet and internal pressure gauge. The vacuum system reaches a working pressure of 1x10-2 mbar. The evaporation technology used is cathodic arc. The system consists of four current sources for activating the four arc cathodes, a current source for the sputtering cathode and a power generator for the bias that is applied to the samples.

To clean the samples before deposition, a glass beaker was filled with special detergent (Extran brand) and shaken with ultrasound for 10 minutes. The procedure was repeated using acetone p.a. Titanium nitride films were then deposited on the M2 and D6 steel substrates with polished surfaces (as described in Item 3.2.1), using a thin layer of intermediate titanium deposited to act as an intermediate layer between TiN and the steels. Before the start of each deposition, the surface of the samples was cleaned using argon plasma for 10 minutes. The intermediate layer of pure titanium, measuring approximately 10 nm, was then deposited. The deposition conditions used are shown in Table 3.1.

3.2.3.2. Depositing   AlN Films and Films

TiN and AlN functionals

The depositions of aluminum nitride films and functional titanium nitride and aluminum nitride films were carried out in a stainless steel 304 vacuum chamber 900 mm high by 1000 mm in diameter, with a gas inlet and internal pressure gauge (Figure 3.3). The vacuum system consists of a mechanical pump and a diffuser pump, and is controlled by an absolute pressure gauge. The substrate temperature is measured by a K-type thermocouple. The vacuum system reaches a working pressure of 1.5 mTorr. The evaporation technology used is magnetron sputtering of metal ions. This magnetron sputtering film growth system was designed and built in the Physics Department of the Instituto Tecnológico de Aeronáutica (IEFF).

FIGURE 3.3. Magnetron sputtering system for depositing AlN films and functional TiN and AlN films.
To clean the samples before deposition, the same procedure was used as described in Section 3.2.3.1, with the exception of cleaning the substrates by sputtering, which is not available in this system. The AlN film was deposited for two hours and measured using a profilometer to obtain a deposition rate. With this rate, it was possible to estimate the time needed to obtain these films with thicknesses of 250 nm. Aluminum nitride films and functional titanium nitride and aluminum nitride films were then deposited on apo M2 and D6 substrates with polished surfaces (according to Item 3.2.1), using a thin layer of intermediate titanium for TiN and intermediate aluminum for AlN, deposited to act as an intermediate layer between the nitride films and the apo. Before the start of each deposition, the target (titanium or pure aluminum) was cleaned using argon plasma for 10 minutes and then the intermediate layer of approximately 10 nm was deposited on the samples. After this procedure, the titanium nitride and aluminium nitride films were deposited. The deposition conditions used for the aluminum nitride films are shown in Table 3.1. The deposition conditions for the TiN and AlN functional films are shown in Table 3.2.

TABLE 3.1. Deposition conditions by cathodic arc for titanium nitride films and by magnetron sputtering for aluminum nitride films.

Film type	Gases (sccm)		Power (W)	Pressure (mTorr)	Temperature (° C)	Deposition time (h)
	Air	N2				
TiN (stoichiometric)	1200	1200	80	7,5	220 e 450	7

AlN (stoichiometric)	23	23	200	1,5	143	2

TABLE 3.2. Magnetron sputtering deposition conditions for titanium nitride and aluminum nitride functional films.

Substrate	Type of film		Pressure (mTorr)	Substrate temperature (° C)	Deposition time (min)	Layers
M2 (AISI M2) and D6 (AISI D6)	TiN	Conditions	2-2,3	177(*)	0,5	1° layer
					15	2° layer
					15	3° layer
					15	4° layer
					15	5° layer
	AlN	Condition2	1,5	114(*)	2,5	1° layer
					15	2° layer
					15	3° layer
					15	4° layer

(continued)

TABLE 3.2 (Conclusion).

					15	5° layer
		Condition3	1,5	114(*)	0,5	1° layer
					15	2° layer
					15	3° layer
					15	4° layer
					15	5° layer
		Condition4	1,5	135(*)	0,58	1° layer
					15	2° layer
					15	3° layer
					15	4° layer
					15	5° layer

(*) temperature coming from the plasma generated in the chamber and without external heating.

3.2.4. Characterization of the Surfaces and Interfaces of the Deposited Films

The surfaces of the deposited films were characterized using analyses of the phases present by X-ray diffraction, topography and microstructure by SEM, chemical characteristics by means of spectral curves by EDX (Energy Dispersive Spectrometry), AFM (Atomic Force Microscopy), electron backscattering, XPS (X-ray Photoelectron Spectroscopy) and RBS (Rutherford Backscattering Spectrometry) analyses. The film-substrate samples were subjected to the 4-point bending test in order to analyze the mechanical behavior of the materials under study and identify the tension at which the film breaks and/or peels off.

In this work, the experimental RBS spectra were analyzed using the program RUMP (Rutherford Universal Manipulation Program), which consists of a series of subroutines in the FORTRAN language, designed for data analysis and simulation of Rutherford backscattering spectra (RBS). This program was developed at Cornell University by Dr. J. M. Mayer's research group and is marketed by Computer Graphic Service[134] .

3.2.4.1. Analysis of crystalline phases by X-ray diffraction

X-ray diffraction analysis was carried out on all the surfaces of the samples with titanium nitride and aluminium nitride films and the intermediate films deposited. The parameters used were the same as those listed in Item 3.2.2.4.

3.2.4.2. Topographical analysis by SEM

The surfaces of all the nitride films deposited were observed and photomicrographs taken at various magnifications.

3.2.4.3. Chemical Analysis of Surfaces by EDX

Characterizations of the chemical elements present were carried out by EDX on all the TiN and AlN films deposited and their respective spectral curves were obtained.

3.2.4.4. Characterization of Film-Substrate Interfaces by EDX

In order to measure and analyze the TiN film-Ti-apo film M2 interfaces and the TiN film-Ti-apo film D6 interfaces by line mapping, the sheet-shaped samples were cut in half in order to obtain the transverse sections of the film-substrate assembly. Cooling with distilled water was used to avoid heating the sample during the procedure, which could alter the thickness of the interdiffusion region. The new surfaces created were manually sanded and polished by sliding the sample parallel to the film surface. The steps of the procedure adopted were carried out as described in Item 3.2.1, both for sanding and polishing, and after using each type of abrasive, the sample was subjected to ultrasonic cleaning.

The analyses were carried out by line mapping (using the EDX detector). Line mapping analysis was carried out on the samples with titanium nitride film with an intermediate titanium layer, with the aim of studying the formation of interfaces between these films and the M2 and D6 steel substrates. These analyses could not be carried out for the other films, as they did not reach an adequate thickness according to the limitations of this technique for very thin films, due to the "pear" effect of the electron beam when in contact with the sample[21]. The TiN films reached an adequate thickness for this analysis because the deposition rate of films deposited in cathodic arc equipment is much higher than the rate of films deposited by magnetron sputtering[70,120,124,125].

3.2.4.5. Analysis by Atomic Force Microscopy (AFM)

The surfaces of the titanium nitride and aluminum nitride films deposited on M2 and D6 steels were analyzed by AFM to check surface roughness.

The AFM equipment used was manufactured by Shimatzu, model SPM-9500J3, and is located in the Physics Department of the Aeronautics Institute of Technology (ITA).

3.2.4.6. Backscattered Electron Analysis of the TiN-M2 and TiN-D6 Interface

The interfaces of the titanium nitride films deposited on M2 and D6 were observed by backscattered electrons. These analyses could not be carried out for the other films either, for the same reason as discussed in Section 3.2.4.4.

3.2.4.7. Analysis by X-ray Photoelectron Spectroscopy (XPS)

The surfaces of the TiN and AlN functional films were analyzed by XPS for quantitative analysis of the elements and chemical compounds present on the surfaces.

The samples were characterized by X-ray photoelectron spectroscopy (XPS), using the surface spectromicroscope (Kratos XSAM HS) available at the Materials Characterization and Development Center (CCDM), installed at the Federal University of Sao Carlos (UFSCar).

The XPS analyses were carried out in an ultra-high vacuum environment (pressure in the range of 10^{-9} Torr). Aluminum Ka radiation was used as the excitation source, with an energy of 1486.6 eV and a power of 168 W, given by a voltage of 14 kV and an emission of 12 mA. The 284.8 eV value for the C 1s photoelectric line associated with the C-C and/or C-H of adventitious hydrocarbons was used as the binding energy reference. The peaks were fitted using the program supplied by the equipment manufacturer, with Gaussian curves (for the C 1s and O 1s lines) and mixed Gaussian/Lorentzian curves (for the other lines), background subtraction using the Shirley method and the least squares routine.

3.2.4.8. Rutherford Backscattering Analysis (RBS)

The surfaces of the titanium nitride and aluminum nitride films deposited on M2 and D6 steels were analyzed by RBS and their respective spectra were obtained to evaluate the ratios between the concentrations of the film's constituents.

The RBS (Rutherford Backscattering Spectrometry) analysis equipment used was assembled by researchers from the Materials and Ion Beam Laboratory of the Physics Institute (LAMFI) at the University of São Paulo

(USP).

3.2.4.9. RUMP Simulation for Rutherford Backscattering (RBS)

Several simulations were carried out using the RUMP (Rutherford Universal Manipulation Program) program for titanium nitride and aluminum nitride films on M2 and D6 steels and their respective spectra were obtained. These, in turn, were compared with the spectra obtained in the previous item.

Because the substrates used in this work have chromium and iron in their chemical composition, which in this case are elements that have higher atomic weights than the elements titanium, aluminum and nitrogen, in this thesis work a study was carried out using RUMP simulations to find better growth parameters and, consequently, a better deposition plan for TiN and AlN films on substrates containing Cr and Fe. With this study, it was possible to optimize the use of RBS for films grown on Cr- and Fe-based substrates.

3.2.4.10. 4-Point Bending Tests

As no standard was found for flexural testing of film substrates, the parameters presented in ASTM 855-90[130] were adapted in this work. The main objective was to determine the tensile strength of the deposited films. The tests were therefore carried out using loads as shown in Figure 3.4.

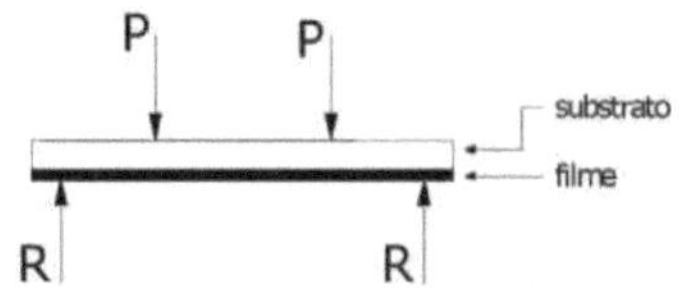

FIGURE 3.4. Schematic drawing showing the
loads applied
to the film substrate in a
4-point
bending test.

Samples of M2 and D6 steel were prepared to the dimensions specified in ASTM E 855 - 90[130] , arriving at the following values:

-   Sample thickness (substrate + film): 0.4mm (according to standard 0.25mm);
-   Sample length: 62.5mm;
-   Sample width: 4mm (according to standard 3.81mm);
-   Bottom clearance of the device: 37.5mm; and -
    Device top gap: 25mm.

Titanium nitride and aluminum nitride films were deposited, 30 mm by 4 mm in the center, on prepared samples of apo M2 and D6 under the same conditions listed in Tables 3.1 and 3.2.

A device was designed and built to the dimensions specified in the ASTM E 855 - 90 standard[130] to carry out the four-point bending tests.

Four-point bending tests were carried out on all the prepared samples.

The test conditions were:

-   displacement: 0.5mm/min;
-   temperature : 27° C; and
-   ambient humidity: 60%.

The equipment used for the bending test was a universal testing machine (INSTRON, model 4301) located in the Materials Division of the Aeronautics and Space Institute of the Aerospace Technical Center (CTA).

CHAPTER 4
PRELIMINARY RESULTS AND DISCUSSION

4.1. Characterization of Substrate Surfaces

4.1.1. Characterization of substrate surfaces by SEM and EDX

In order to study the surfaces of the steels before the films were deposited, the surfaces of the M2 and D6 steel substrates were prepared according to the procedures adopted in Item 3.2.1, and are shown in Figures 4.1 and 4.2, respectively.

The surface of an M2 steel sample shows lighter spots distributed evenly, but varying in size, and the presence of some black dots (Figures 4.1a and 4.1b). EDX analysis shows that the light regions have a higher concentration of molybdenum, tungsten and vanadium (Figure 4.1c), compared to the spectrum obtained for the main region of the surface, which EDX analysis also showed elements characteristic of the substrate's chemical composition (Figure 4.1d).

Topographical analysis of apo D6 as seen in Figure 4.2 shows a fairly homogeneous surface with a considerable number of defects in the form of shallow scratches and pores of varying sizes (Figures 4.2a and b). These observations of the surface of this apo also show the presence of darker, non-uniformly distributed spots. EDX analysis shows that these dark spots have a higher concentration of chromium in their composition (Figure 4.2c), compared to the rest of the surface of this substrate, which was also analyzed by EDX (Figure 4.2d).

The next section discusses the RBS analyses used to evaluate the atomic concentrations of the chemical components of the M2 and D6 apos used in this thesis.

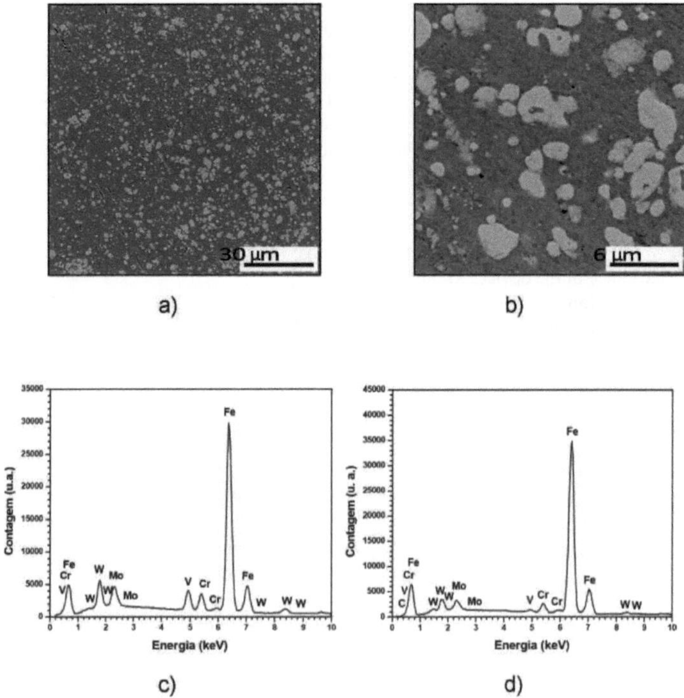

FIGURE 4 .1 a) and b) SEM photomicrographs of the surface of the M2 steel substrate, c) EDX spectrum of the clear regions and d) EDX spectrum for the surface of the M2 steel.

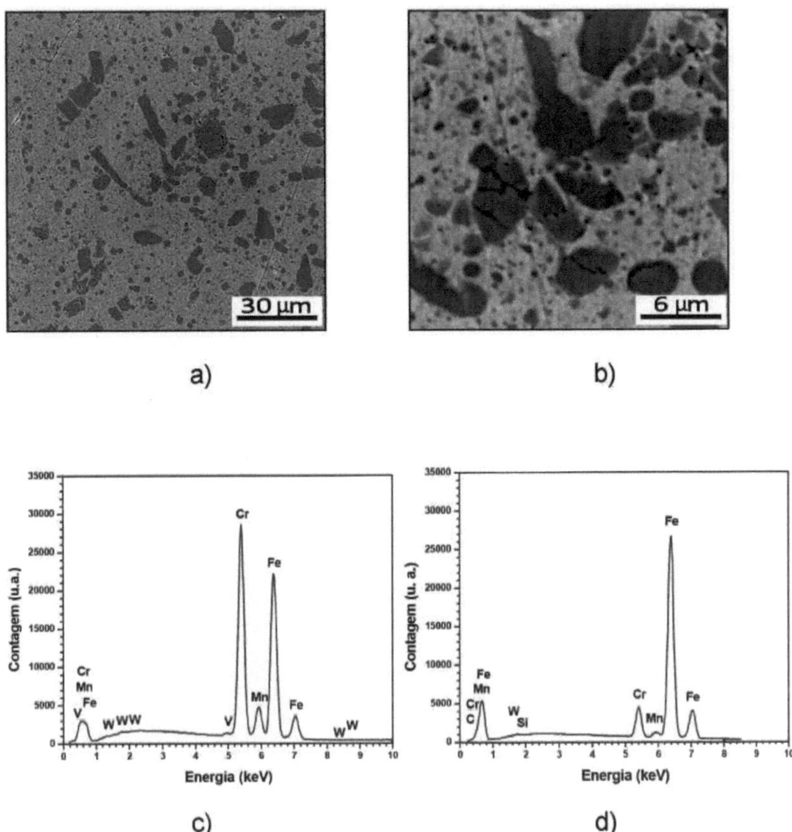

FIGURE 4 .2 a) and b) SEM photomicrographs of the surface of the D6 steel substrate, c) EDX spectrum of the dark regions and d) spectral curve of the D6 steel surface.

4.1.2. Substrate surface analysis by RBS

To determine the atomic composition of M2 and D6, chemical analyses were carried out using atomic absorption spectrometry, gravimetry and the LECO C and Si determiner.

The chemical analyses obtained for apo M2 showed the following chemical composition: 0.92% C, 3.99% Cr, 5.15% Mo, 2.01% W, 6.24% V and 81.69% Fe. These values are very close to those presented by the manufacturer (as described in Item 3.1).

The results of the chemical analysis carried out on apo D6 showed chemical composition values of: 1.93% C, 0.25% Si, 0.26% Mn, 11.00% Cr, (not detected) W and 86.56% Fe. These values are also in line with those presented by the manufacturer, with the exception of tungsten, which was not detected. In this work, these chemical analysis values are used because they are more reliable than those presented by the manufacturer.

4.2. Characterization of the deposited films

4.2.1. Titanium Nitride Films

4.2.1.1. Characterization by X-ray diffraction

The films deposited on the surfaces of M2 and D6 steels, at temperatures of 220 and 450° C, were characterized by conventional X-ray diffraction. Figure 4.3 shows the X-ray diffractograms of the M2 steel substrates and the films deposited on the M2 steel, confirming the presence of TiN.

These X-ray diffractograms also show the presence of the ferrite phase and the M6C phase, which are components of the steel substrate. According to X-ray diffraction analyses carried out by other authors[135] , the M6C phase is composed, in atomic percentage, of 45.70% Fe, 23.40% Mo, 19.30% W, 5.20% Cr and 6.40% V. It can be seen that, under the deposition conditions used, the nucleation and growth of TiN crystals occurs preferentially in the (111) direction. The presence of metallic titanium in the deposited intermediate film was not observed. The penetration of the X-ray beam is of the order of 1000 nm in the sample to be analyzed and the intermediate film deposited with a thickness of 10 nm becomes imperceptible to the X-ray detector.

As can be seen in Figure 4.3, the apo M2 samples with TiN films deposited at 450° C show the presence of TiN(200) and TiN(111), indicating the nucleation and growth of TiN crystals in two crystalline directions preferentially. This fact proves that, as the temperature increases, while maintaining the other deposition conditions, preferential growth begins to occur in the (111) and (200) directions.

The individual diffraction peaks of the a-iron (ferrite) of apo M2 and apo M2 modified by the presence of the titanium nitride film are shown in Figure 4.4.

As a result of the analysis of these diffraction peaks, it is suggested that there has been a decrease in the density of the crystalline planes of iron, due to a decrease in the atomic density of these planes due to the presence of Ti and N atoms. The changes in the relative positions of the peaks are not sufficient to determine a variation in the interplanar distances of the (110) plane family of a-iron. This is due to the small thickness of the interface formed by thermally activated diffusion.

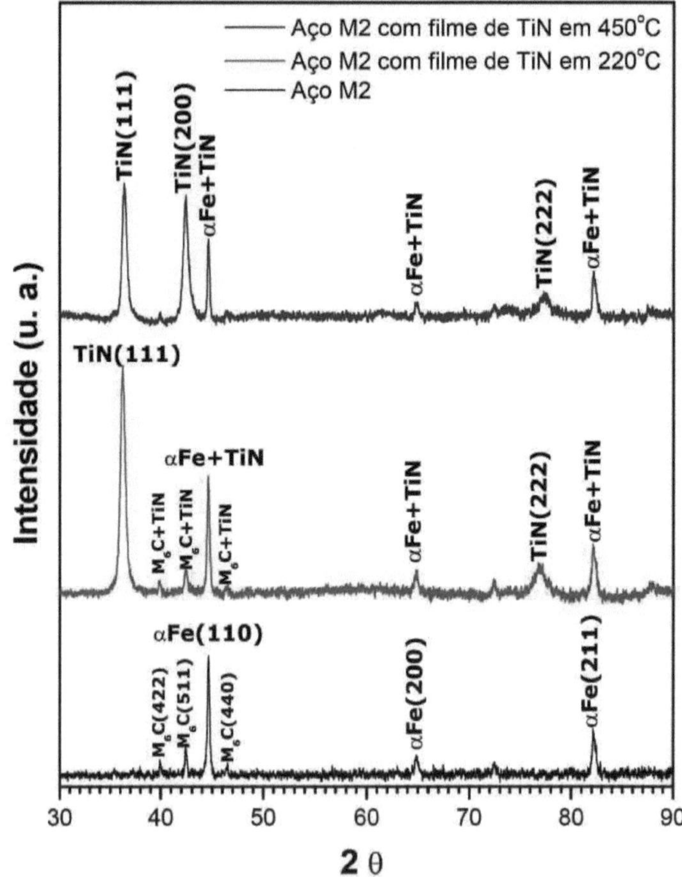

FIGURE 4 .3 X-ray diffractograms for the apo M2 substrate and for the titanium nitride films deposited on apo M2 at temperatures of 220 and 450 C.°

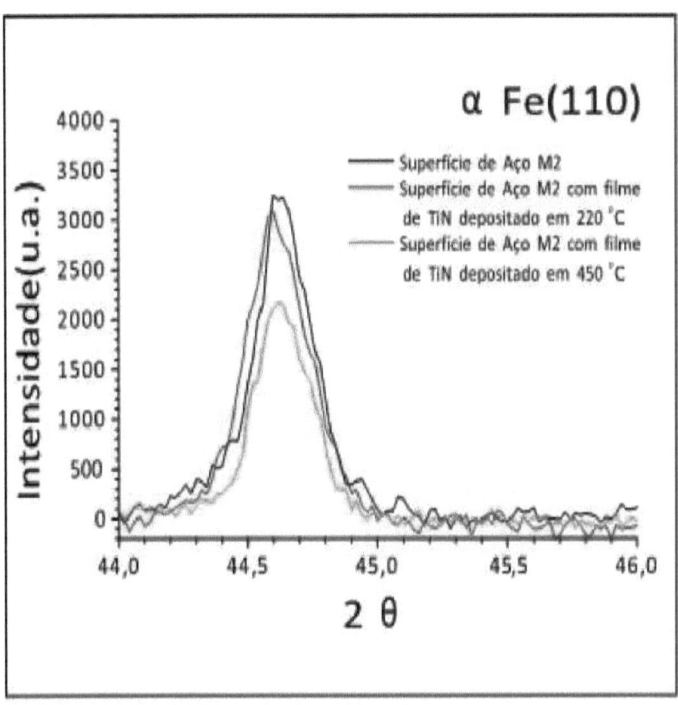

FIGURE 4.4 Individual X-ray diffraction peaks of a-iron (ferrite) from M2 steel and M2 steel modified by the presence of titanium nitride film deposited at temperatures of 220 and 450 C.°

Although the composition of this interface varies depending on the depth, the diffraction result corresponds to an average of the intensities of the diffracted X-ray beam and the interplanar distances of the crystal planes responsible for the diffraction.

The individual diffraction peaks of the a-iron (ferrite) component of D6 steel and D6 steel modified by the presence of the titanium nitride film deposited under the conditions specified in Table 3.1 are shown in Figures 4.5 and 4.6.

As can be seen, the D6 steel samples with TiN films deposited at temperatures of 220 and 450° C, show growth characteristics similar to the films deposited on M2 steel substrates, i.e. they show the peaks TiN(111) for deposits at 220° C and TiN(200) and TiN(111) for deposits at 450° C. This result proves that the deposition temperature affected the nucleation and growth of the TiN films, but the type of substrate apo does not seem to have affected these characteristics. Both have alpha iron and, in particular, apo M2 has another crystalline phase. This may have affected the interplanar distances, but the diffractograms did not detect this due to the lack of resolution. On the other hand, an expert in the field said that it was impossible to analyze the interplanar distances using high-resolution X-ray diffraction because it was a polycrystalline structure. Analyzing a polycrystalline structure using high-resolution X-ray diffraction would involve many variables that would be difficult to control. Another fact may be related to stress relief and recrystallization, which should be greater for the higher deposition temperature because, due to the forming process, these anneals are possibly stressed. On the other hand, they also have a brittle, hard and highly resistant martensitic structure.

The individual diffraction peaks of the a-iron (ferrite) of apo D6 and of apo D6 modified by the presence of the titanium nitride film are shown in Figure 4.6, whose behavior is similar to that of apo M2 with and without the TiN film deposited.

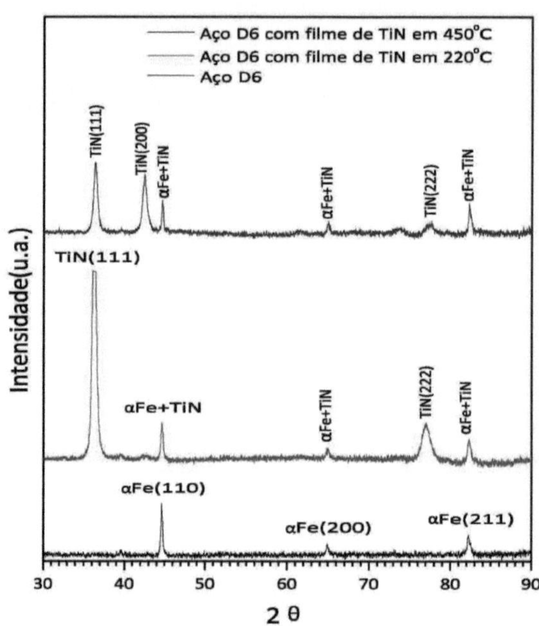

FIGURE 4 .5 X-ray diffraction curves for apo D6 and titanium nitride films deposited at temperatures of 220 and 450° C on apo D6.

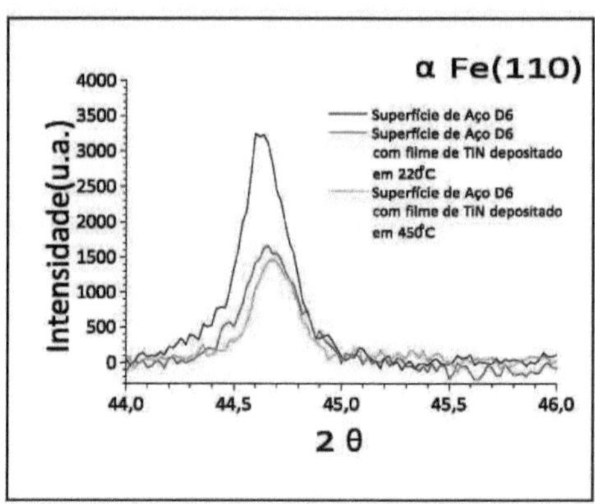

FIGURE 4.6 Individual X-ray diffraction peaks of iron a (ferrite) from apo D6 and apo D6 modified by the presence of titanium nitride film, deposited at temperatures of 220 and 450 C.°

4.2.1.2. Characterization by SEM and EDX

Intermediate titanium films and titanium nitride films were deposited on the M2 and D6 substrates at temperatures of 220 and 450 C.°

The titanium nitride films deposited on the apo M2 substrates at a temperature of 220° C were porous and had a homogeneous distribution of pore sizes. Figures 4.7a, b and c show the microstructures of the surface of a titanium nitride film deposited on an apo M2 substrate, which is representative of the samples obtained. The spectral curve obtained by EDX for this surface is shown in Figure 4.7d, proving the presence of the component elements of the film and the substrate.

Some authors relate the microstructure of dense or porous TiN films to parameters of the deposition process, combining the power used for sputtering titanium ions with the bias applied to the substrates (to aid the thermally activated diffusion process)[62] .

Analysis of the titanium/apo interface in these samples using EDX (line mapping) was not possible because the thickness was too small (10 nm) to be detected by this technique[21] .

To study the nitride/apo interface in these samples of TiN films deposited on apo M2 at 220° C, via atomic diffusion at the interface, analyses were carried out using line mapping (EDX) of the cross sections of the film/substrate assembly. As can be seen in Figure 4.8, the curves show a marked variation in the concentration of titanium and substrate components in the interface region. The curve corresponding to nitrogen in the film region shows concentrations well below those expected. This is characteristic of semi-quantitative EDX analyses for low atomic weight chemical elements. The localized increase in chromium concentration in the substrate must be related to the segregation of this element in the graphite contours.

However, this mapping technique is not entirely punctual because the characteristic X-rays are pear-shaped. Therefore, these analyses are considered semi-quantitative[21] . Thus, the actual value for the existing interface region is slightly smaller than that shown by the curve, which is approximately 0.8 pm.

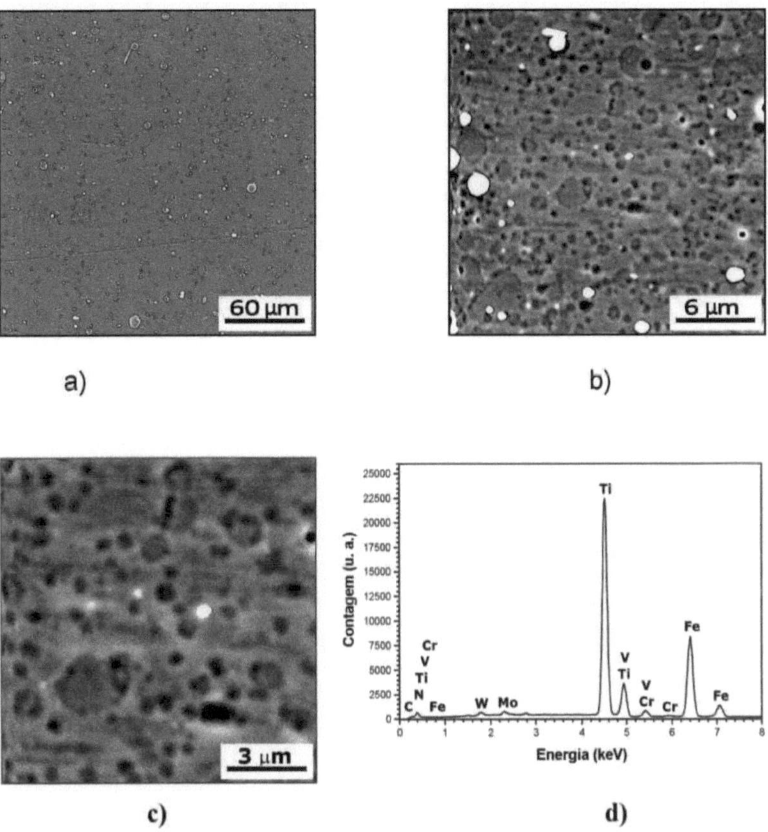

FIGURE 4 .7 Photomicrographs obtained in SEM for the TiN film deposited on the surface of the M2 steel substrate at a temperature of 220° C, a) 500x, b) 5000x and c) 10000x and d) spectral curve obtained by EDX of the surface of this film.

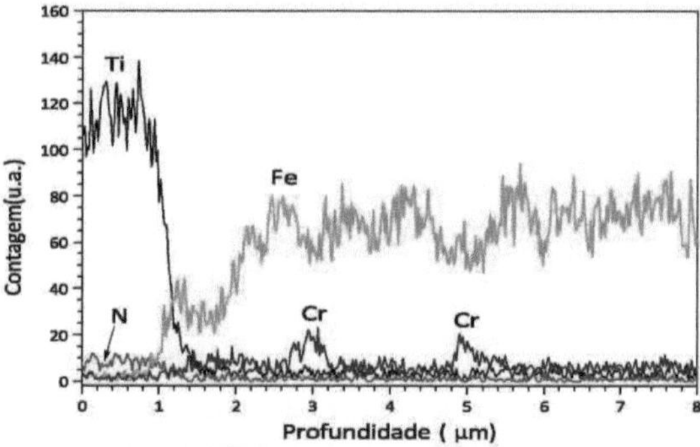

FIGURE 4 .8 Relative concentration profile curves of the chemical elements present in the TiN film sample deposited on apo M2 at 220° C, obtained by EDX.

The titanium nitride film deposited on the M2 steel sample at a temperature of 450° C had a more homogeneous surface (Figures 4.9a, b and c). The surface of the film is porous and has a homogeneous distribution of pore sizes. Figure 4.9d shows the EDX spectral curve obtained for this surface, proving the presence of the component elements of the film and the substrate.

The interfaces of these TiN films, deposited on M2 steel at 450° C, via atomic diffusion at the interface, were analyzed using line mapping of the cross-sections of the film/substrate assembly. As can be seen in Figure 4.10, the curves show a less marked variation in the concentration of the film and substrate components when compared to the curves of the film deposited at 220° C. Thus, the behavior shown by these curves indicates the formation of a more diluted interface at the TiN/substrate junction. This diffusion must have been facilitated by the higher temperature at which the film was deposited in this case, i.e. at 450° C. This result suggests that a solid solution formed at this interface, with a gradual variation in the composition of the interface.

The line mapping is consistent with an increase in the interface region with temperature.

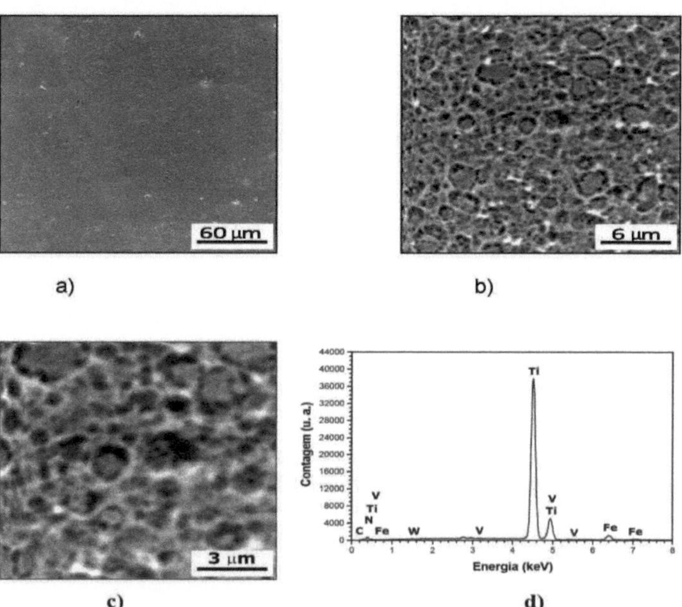

FIGURE 4 .9 Photomicrographs for the TiN film deposited at 450oC on the surface of apo M2 obtained by

SEM, a) 500x, b) 5000x and c) 10000x and d) EDX spectral curve of the surface of this film.

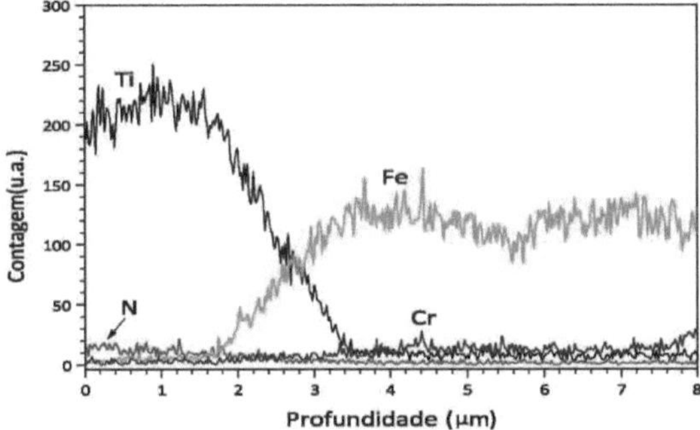

FIGURE 4 .10 Profile curves, obtained by EDX, of the relative concentration of the chemical elements present in the apo M2 substrate and in the TiN film deposited at 450° C on apo M2.

The titanium nitride film deposited on the apo D6 substrate at a temperature of 220° C was porous and had an inhomogeneous distribution of pore sizes (Figures 4.11a, b and c). Figure 4.11b shows the presence of large pores. However, this film has a denser microstructure than those shown above. The presence of white dots can also be seen, and EDX analysis has shown chemical elements similar to the other regions of the surface, i.e. component elements of the TiN film. In this case, the white dots would be TiN agglomerates formed on the surface during deposition. In an SEM image analysis, when it is not possible to focus on the uppermost regions of a given sample, which in this case have an interaction with the electron beam, these regions usually show an intense white color. Figure 4.11d shows the presence of the component elements of the film and substrate, obtained by EDX for this surface.

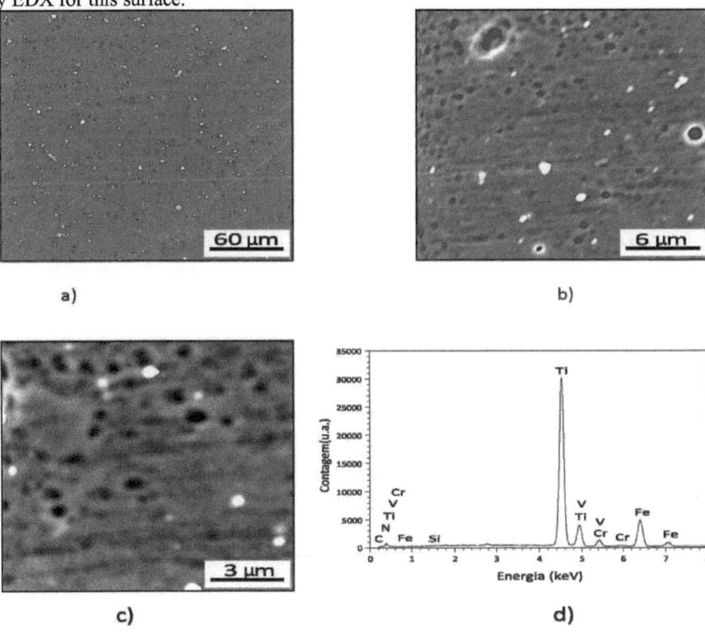

FIGURE 4 .11 Photomicrographs of the TiN film deposited on the surface of apo D6 at a temperature of 220° C obtained by SEM, a) 500x, b) 5000x and c) 10000x and d) spectral curve of the surface of this film obtained by EDX. surface of this film obtained by EDX.

Analyses of these TiN films, deposited on apo D6 at 220° C, via atomic diffusion at the interface, were carried out using line mapping (EDX) of the transverse sections of the film/substrate assembly. The behavior shown by the line mapping curves (Figure 4.12), compared with the curves shown by the TiN film deposited on apo M2 at the same temperature of 220° C (Figure

4.8) The concentration of the film and substrate components varies less markedly.

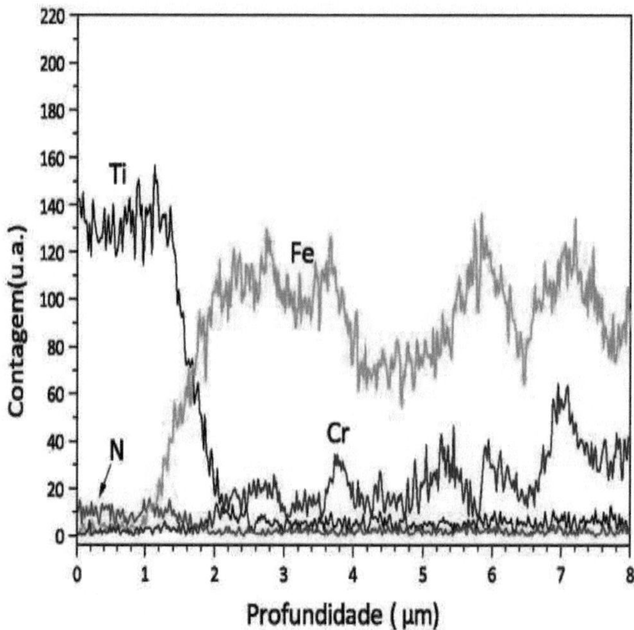

FIGURE 4 .12 Curves obtained by EDX of the relative concentration profiles of the chemical elements present in the apo D6 substrate and in the TiN film deposited at 220° C on apo D6.

The titanium nitride film deposited on the D6 steel sample, at a temperature of 450° C, shows the occurrence of detachment of the layer closest to the surface (Figures 4.13a, b and c). It can be seen in the remaining parts that these layers are much denser than the inside of the film. This difference in porosity may have been responsible for the detachment due to thermal shock during the relatively rapid cooling process. Spot analyses of these denser parts using EDX showed spectral curves similar to those of the porous region of the surface. The spectral curve

of the entire surface, obtained by EDX, is shown in Figure 4.13d, proving the presence of the component elements of the film and the substrate.

Interface analysis of these TiN films, deposited on D6 steel at 450° C, via atomic diffusion at the interface, was carried out using line mapping (EDX) of the transverse sections of the film/substrate assembly. As shown in Figure 4.14, the curves also show a less marked variation in the concentration of the film and substrate components compared to the film obtained at 220° C. This result confirms the formation of a solid solution at this interface, with a gradual variation in the composition of this interface.

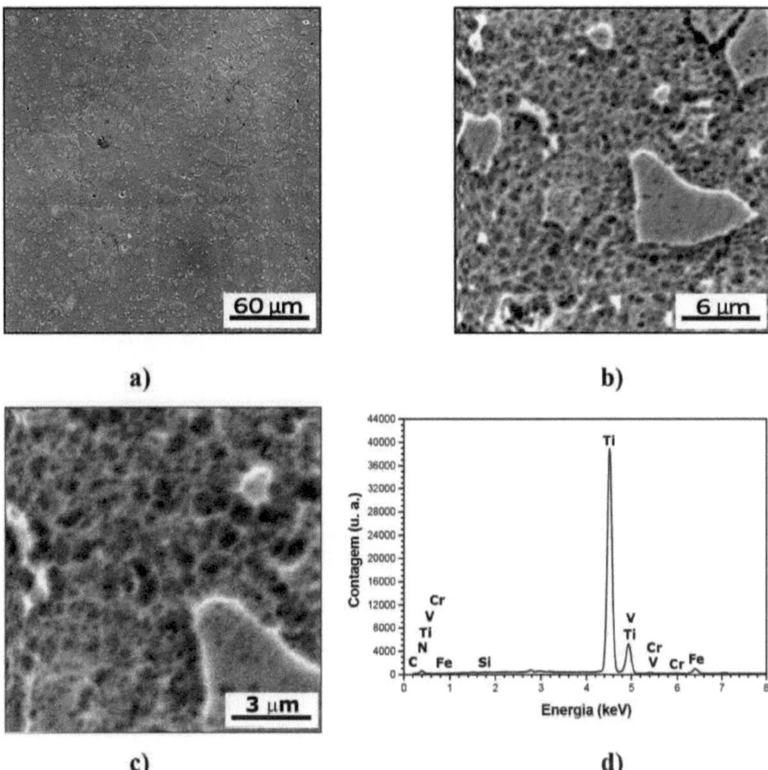

FIGURE 4 .13 Photomicrographs for the TiN film deposited on the surface of D6 steel at a temperature of 450° C, obtained in SEM, a) 500x, b) 5000x and c) 10000x and d) EDX spectrum of the surface of this film. The behavior of these curves showed a less marked variation than the curves obtained for the TiN film deposited on apo M2 at the same temperature (Figure 4.10). This indicates the formation of a thicker TiN/apo interface for the film deposited on the apo D6 substrate.

These results show that the films deposited on both apo substrates at the lowest temperature have a denser microstructure and a more homogeneous distribution of pores. However, TiN films deposited on D6 substrates have a denser microstructure than those deposited on M2 apo. On the other hand, these relative atomic concentration profile curves, when compared, since they were obtained using the same technique with its appropriate (semi-quantitative) limitations[21,99,103-105] , show that the interface was thinner when the depositions were made at a temperature of 220° C. Therefore, for the ideal deposition conditions for these films, it is necessary to choose a sufficiently high temperature (= 450° C) for a larger interface region, but which at the same time does not cause the layer closest to the surface to peel off.

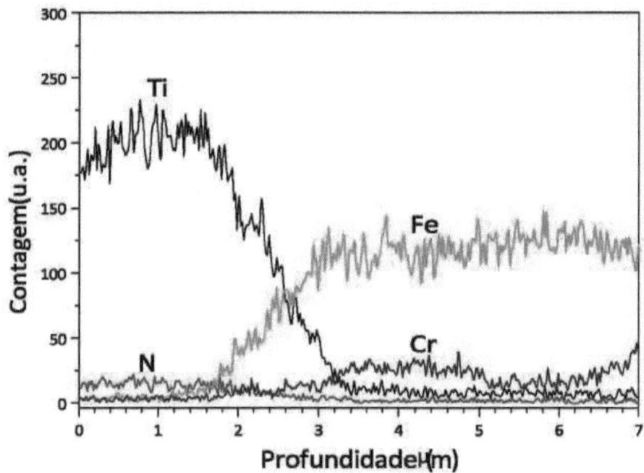

FIGURE 4 .14 Relative atomic concentration profile curves, obtained by EDX, of the chemical elements present in the TiN film deposited on apo D6 at 450° C and on the apo D6 substrate.

4.2.1.3. AFM characterization

The surface roughness of the titanium nitride films deposited at 220 and 450° C was mapped using the atomic force microscopy (AFM) technique. The images resulting from these mappings of the surfaces of the M2 high speed steel and D6 tool steel samples with titanium nitride films deposited at 220 and 450° C are shown in Figures 4.15 and 4.16.

Observations of the surface morphology obtained by AFM indicate that the TiN film deposited on M2 high speed steel at 220° C is slightly less rough on the surface than the TiN film deposited at 450° C, with roughness values of 9.458 nm and 9.585 nm, respectively (Figure 4.15). The titanium nitride surface deposited at 220° C showed a heterogeneous microstructure with nanograins of various sizes, although with similar shapes (Figure 4.15a). This characteristic is the result of deposition at lower temperatures, where the coalescence of the nanogrids has not yet been fully completed.

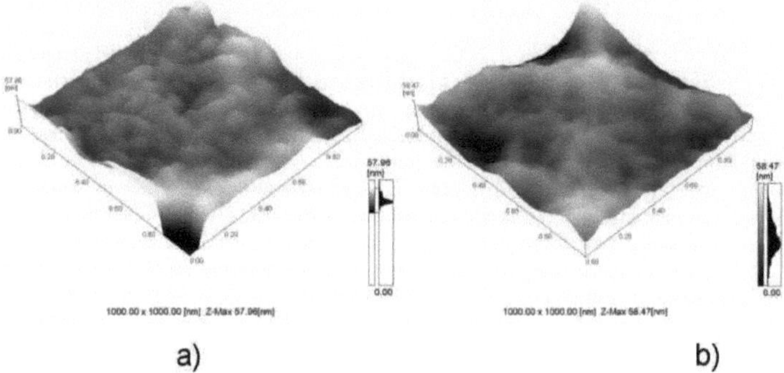

FIGURE 4 .15 AFM images of the surfaces of titanium nitride films deposited on M2 steel at temperatures of: a) 220° C and b) 450 C.°

AFM observations of the surface morphology of the TiN film deposited on D6 tool steel at 220° C also showed that it had lower surface roughness than the TiN film at 450° C, with roughness values of 3.988 nm and 7.438 nm, respectively (Figure 4.16). The titanium nitride surface deposited at 220° C showed a much more homogeneous microstructure with nanograins of similar size and shape, but less coalesced (Figure 4.16a), when compared to the film deposited on the M2 steel substrate (Figure 4.15a).

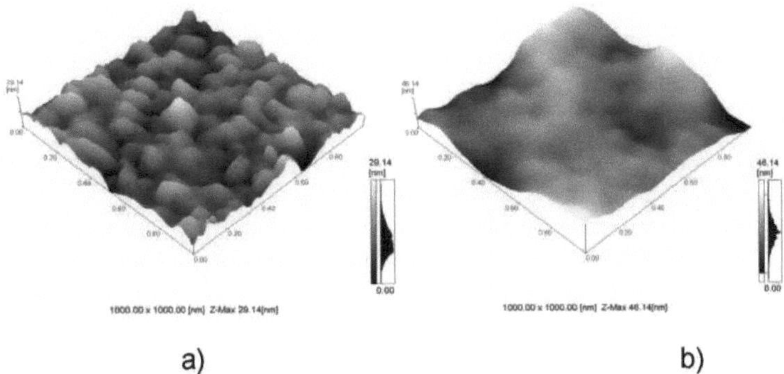

a)   b)

FIGURE 4 .16 AFM images of the surfaces of titanium nitride films deposited on D6 steel at temperatures of: a) 220° C and b) 450 C.°

The films deposited at a temperature of 220° C had a nanostructure. At this temperature, there was adsorption and surface diffusion of the ions and/or atoms on the surface of the substrates, with enough time for them to settle and form in a nanostructured and uniform manner. The balance between adsorption and desorption of ions and/or atoms at this temperature is ideal for this type of deposited film. The films deposited at 450° C formed very quickly and with little structure on the surface of the substrates, as the amount of ions and/or atoms adsorbed was much greater than those desorbed from the substrate surface. This greater quantity of ions and/or atoms adsorbed on the surface of the substrates was due to the greater energy coming from the deposition temperature, which meant that the accommodation and arrangement of these ions and/or atoms was very rapid. The type of substrate and film and the control of film growth parameters, such as temperature, pressure, etc., are of great importance in the balance between the ions and/or atoms arriving and leaving the surface of the substrates during the deposition process[ ].[125,136]

4.2.1.4. Characterized by Backscattered Electron Microscopy

Analyses of the TiN/apo interfaces in these samples, using electron backscattering, are shown in Figures 4.17 and 4.18. In the images, near the interface region, there are no signs of the presence of the intermediate titanium layer. Although the backscattered electron detector detected the presence of this layer at the interface, which is of the order of 10 nm, its backscattered image response was unable to distinguish it from the TiN film, which in this case has a very close atomic weight. As can be seen in Figure 4.17, the TiN film deposited on apo D6 at a temperature of 220° C has a higher deposition rate and, consequently, a greater thickness than the same film, deposited under the same conditions, on the surface of apo M2. This difference in thickness may be related to a difference in energy on the surface of each apo, which favored greater growth of this film on apo D6. The white spots shown in Figure 4.17a come from a phase with a higher concentration of molybdenum and tungsten, as discussed in Item 4.1.1. These elements have a higher atomic weight than the other elements and, according to the electron backscattering technique, are shown in the image with a lighter color[100,103] .

The film deposited on apo D6 at a temperature of 450° C also showed a higher deposition rate than the same film, under the same conditions, deposited on the surface of apo M2 (Figure 4.18). However, as can be seen, the samples show small signs of microcracks near the TiN film/apo substrate interface. The interface between the film and the D6 apo substrate showed the presence of small pores. Unfortunately, due to the resolution limitations of the SEM used in this work, an image greater than 10000X magnification was not possible to better visualize these microcracks. These results confirm those already presented in Items 4.2.1.2 and 4.2.1.3, showing better deposition of these films at low temperatures, which is very suitable for the production of the interfaces proposed in this thesis. This behavior may be associated with possible changes in the microstructure of apo M2 and D6 and/or differential thermal shrinkage between the film and the substrate when they are heated to higher temperatures. The refinement in the microstructure of apo D6 observed at both deposition temperatures may be related to the deposition time these substrates were subjected to, which was 7 hours, as described in Table 3.1. Although the manufacturer's manual[133] states that these temperatures would not be sufficient for any changes to occur in the microstructure of these substrates, the deposition time may have been the main reason for this refinement. For

On the other hand, EDX analyses of this cross-section, measured from the region close to the surface, showed a high concentration of nitrogen which decreased throughout its depth.

In both steels, an increase in the thickness of the TiN film deposited was observed as the temperature rose. As already discussed in Section 4.2.1.3, the increase in energy from the deposition temperature makes accommodation and formation of the ions and/or atoms arriving on the substrate surface much faster, considerably reducing desorption.

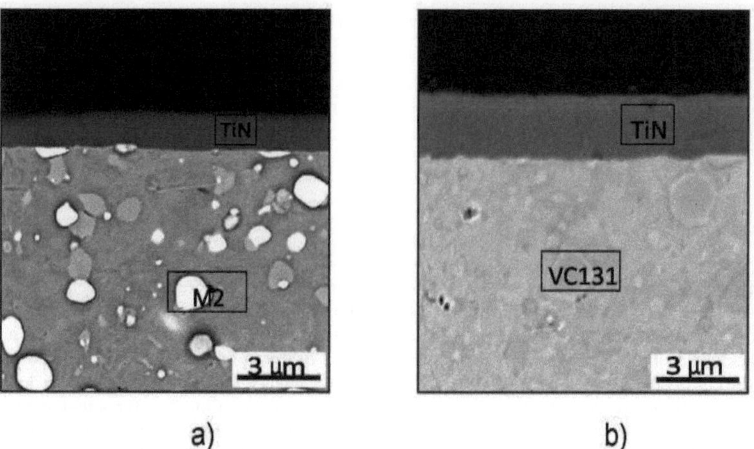

FIGURE 4 17 Images obtained by electron backscattering in SEM, for the cross section of titanium nitride films deposited at a temperature of 220° C on the surface of steels:
a) fast M2 and b) tool D6.

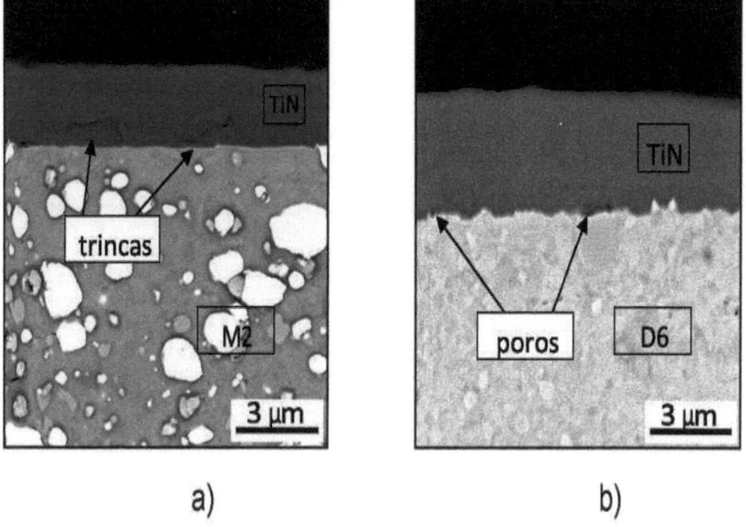

FIGURE 4 18 Cross-sectional images of the titanium nitride films deposited at 450° C, obtained by electron backscattering, on the surface of the steels: a) fast M2 and b) tool D6.

According to the results shown in Table 4.1, the deposition conditions influenced the quality and thickness of the TiN films deposited on M2 and D6 steels. The TiN films deposited at a lower temperature presented a better deposition condition.

TABLE 4.1. Deposition conditions for TiN films and their respective thicknesses.

Type of film	Deposit condition			Thickness (nm)	Defects
	Substrate	Temperature (° C)	Deposition time (h)		
TiN	M2	220	7	1200	-
	D6	220	7	2100	-
	M2	450	7	2400	yes
	D6	450	7	3000	yes

4.2.1.5. Characterized by RBS

4.2.1.5.1. Experimental and Simulated Conventional Data

To determine the atomic composition and atomic concentration profile in depth of the titanium nitride films deposited at temperatures of 220 and 450° C, analyses were carried out using the Rutherford backscattering spectroscopy (RBS) technique. The spectra resulting from the RBS analyses for the samples of M2 quick-release and D6 tool steel with titanium nitride film deposited at 220 and 450° C are shown in Figures 4.19 to 4.22. In these spectra, the black curve represents the experimental values, while the red curve represents the simulation for TiN on M2 and D6 using the RUMP program[ ].[134]

The spectrum obtained by RBS for the titanium nitride film deposited at a temperature of 220° C on M2 steel is shown in Figure 4.19. The concentration values in the TiN layer were obtained by simulation (Table 4.2). The Ti and N concentrations are very close, characterizing a good stoichiometry of the deposited film. There is an amount of carbon in this film, which is possibly due to the use of a diffusion pump. The small concentration of argon may be associated with the cleaning that was carried out on the substrate by sputtering (as described in Item 3.2.3.1.). It was not possible to analyze the intermediate titanium layer due to the overlapping of the Fe peak, which hindered this measurement. The values obtained for the thickness of the TiN film in the simulation were approximately 1500 nm and the concentration of titanium and nitrogen was 36.02%/42.13%, respectively. This thickness is close to that measured by electron backscattering in the SEM (Table 4.1). Analyzing the entire TiN film/Ti film/M2 substrate system at 220° C shows the presence of the following elements: 26.27% N, 16.56% Ti, 15.53% Ar, 5.57% Cr, 28.64% Fe and 7.43% C. The presence of chromium and iron hindered a more precise analysis of titanium and nitrogen (the main components of the TiN film). As chromium and iron have higher atomic weights, they end up overlapping the characteristic peaks of titanium and nitrogen, making it difficult to analyze the TiN film. A further

A detailed analysis, using RUMP, of the best parameters for RBS analysis to minimize this effect is presented in the next section.

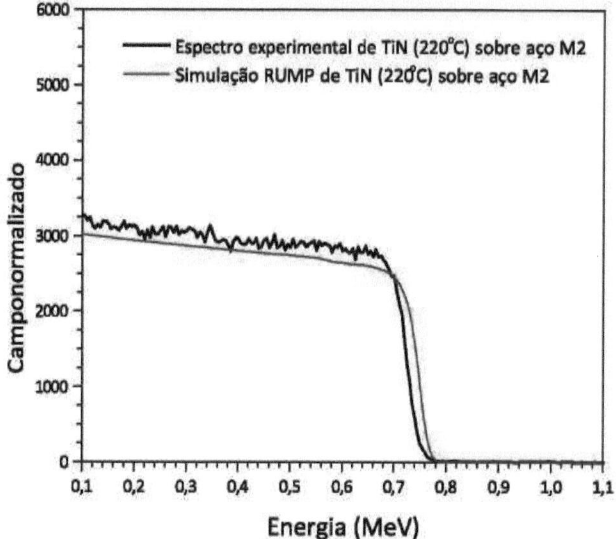

FIGURE 4 19 Experimental and simulated RBS spectra for the TiN film deposited at 220° C on apo M2.
TABLE 4.2. Atomic concentrations of the chemical elements in the TiN film layer deposited under apo M2 at 220° C (deposition conditions specified in Table 3.1).

Ti / N layer (%)	Ti (%)	Air (%)	N (%)	C (%)	Thickness (nm)
50 / 50	36,02	16,26	42,13	5,58	1500
M2 Fast support substrate					

The spectrum curve acquired by RBS for the titanium nitride film deposited at a temperature of 450° C on apo M2 is shown in Figure 4.20. Table 4.3 shows the concentration values in the layer obtained (values obtained by simulation). The titanium element has a similar concentration to nitrogen in this film. This layer also showed a concentration of carbon and a small concentration of argon. The presence of the following elements was detected: 29.89% N, 16.48% Ti, 14.26% Ar, 5.45% Cr, 26.36% Fe and 7.56% C in the TiN film/Ti film/M2 substrate system at 450 C.°

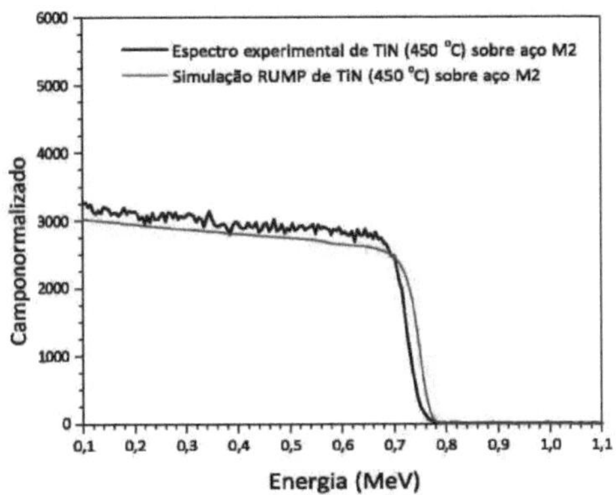

FIGURE 4 20 Experimental spectra obtained by RBS and simulated spectra obtained by RUMP, for the TiN film deposited at 450° C on apo M2.

TABLE 4.3. Atomic concentrations of the constituent elements of the TiN film layer deposited at a temperature of 450° C on apo M2 (deposition conditions specified in Table 3.1).

Ti / N layer (%)	Ti (%)	Air (%)	N (%)	C (%)	Thickness (nm)
50 / 50	34,86	14,24	42,61	8,28	3000
M2 Quick-set substrate					

The values obtained in the simulation were approximately 3000 nm for the thickness of the TiN film and titanium and nitrogen concentrations of 34.86%/42.61%, respectively. The thickness value found is close to that found by electron backscattering (Table 4.1).

The experimental analysis by RBS and the RUMP simulation for the titanium nitride film deposited at a temperature of 220° C on D6 steel are shown in Figure 4.21. When the RUMP simulation was carried out, the concentration values in the layer obtained were obtained (Table 4.4). The concentrations of titanium and nitrogen are close (Table 4.4), showing a very good stoichiometry of the deposited film.

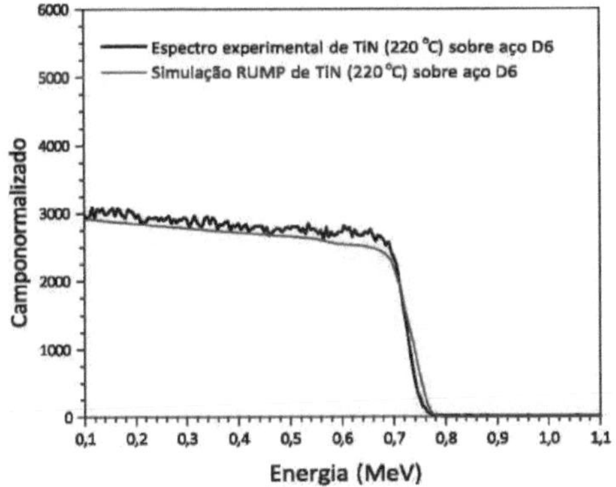

FIGURE 4 21 Spectra obtained by RBS and RUMP for the TiN film deposited at 220° C on D6 steel.

TABLE 4.4. Atomic concentrations of the chemical elements in the TiN film layer deposited (deposition conditions specified in Table 3.1) on apo D6 at a temperature of 220 C.°

Ti / N layer (%)	Ti (%)	Air (%)	N (%)	C (%)	Thickness (nm)
50 / 50	34,22	15,37	41,98	8,42	1800
D6 tool steel substrate					

This film also showed a concentration of carbon and a small presence of argon. A simulation of the TiN film/Ti film/D6 substrate system showed the presence of the following elements: 26.47% N, 22.09% Ti, 11.64% Ar, 5.74% Cr, 25.89% Fe and 8.17% C. The values obtained in the simulation were approximately 1800 nm for the thickness of the TiN film and a titanium and nitrogen concentration of 34.22%/41.98%, respectively. The thickness found in this simulation is approximately the same as that found by electron backscattering (Table 4.1).

The spectrum obtained by RBS for the titanium nitride film deposited at a temperature of 450° C on D6 steel is shown in Figure 4.22. By RUMP, the concentration values in the layer obtained were obtained (Table 4.5). The observed concentration values of titanium and nitrogen are close (Table 4.5), with good stoichiometry of this deposited film. This film also showed a concentration of carbon and a small presence of argon. Simulating the TiN film/Ti film/D6 substrate system shows the presence of the following elements: 23.18% N, 22.97% Ti, 13.41% Ar, 6.52% Cr, 27.65% Fe and 6.27% C. The values obtained in the simulation were approximately 3000 nm for the thickness of the TiN film and a titanium and nitrogen concentration of 33.33%/41.87%, respectively. This thickness is in line with the value already determined by electron backscattering (Table 4.1).

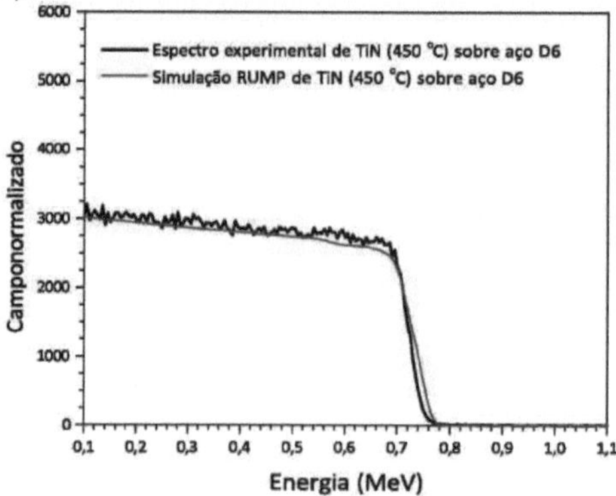

FIGURE 4 22 Experimental and simulated spectra obtained for the TiN film deposited on apo D6 at a temperature of 450 C.°

Despite the difficulty in analyzing titanium and nitrogen by RBS, since chromium and iron (which have higher atomic weights) were present, it was possible to obtain their respective values in atomic concentration in the layer deposited on the substrates. In the analysis of the entire TiN film/Ti film/apo substrate system, a higher concentration of nitrogen was found in the films deposited on apo substrates M2 and D6 compared to the titanium concentration. This may be associated with the fact that the surface of these substrates reacts more easily with this element at the beginning and/or during the deposition of these films. The N and Ti values were very close, showing a good stoichiometry of the film deposited on these substrates. All the films showed a concentration of carbon which possibly came from the diffuser pump used in this deposition. They also showed a small amount of argon which came from sputtering before depositing the TiN film. It was not possible to detect the presence of the intermediate titanium film due to the overlapping of the Fe peak.

TABLE 4.5. Component atomic concentrations of the TiN film layer deposited on apo D6 at 450° C (deposition conditions specified in Table 3.1).

Ti / N layer (%)	Ti (%)	Air (%)	N (%)	C (%)	Thickness (nm)
50 / 50	33,33	15,67	41,87	9,12	3000
D6 tool support substrate					

4.2.1.5.2. RUMP Simulation of TiN Films on M2 and D6 Steels

The values of the conventional parameters used in RBS analyses are: energy of 2.2 MeV and 0 = -7° . These were the same as those used in the analyses presented in Item 4.2.1.5.1. and which showed the overlapping of the peaks of elements with higher atomic weights than the component elements of the TiN film, making a more precise analysis difficult.

Several simulations were carried out in order to find the best parameters for analyzing only titanium and nitrogen in the TiN film. After   many attempts   , and
considering a titanium and nitrogen concentration model of 50%/50%, respectively, 100% Ti as an intermediate layer and substrate compositions M2 and D6 as described in Item 4.1.2, ideal values for energy and 0 were found during RBS analysis of 1.1 MeV and -60° , respectively (Figures 4.23 and 4.24).

Figure 4.23 shows a simulation spectrum of a TiN film deposited on M2 steel. It can be seen that, using the new values for the energy and 0 parameters, it is possible to analyze titanium and nitrogen without interference from the presence of
heavier elements. The arrows indicate the start of detection of each element in the simulated RBS spectrum.

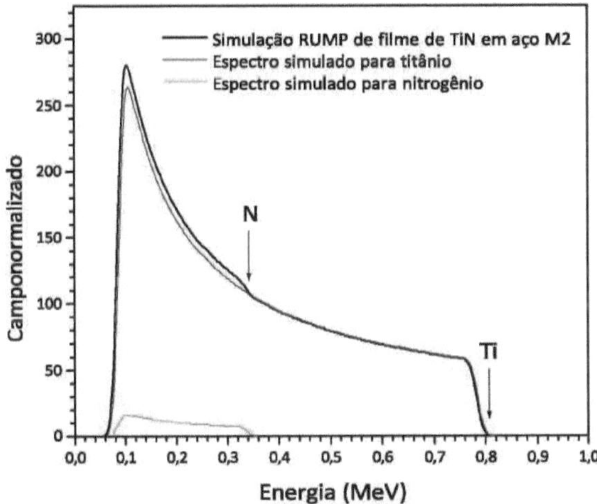

FIGURE 4.23. Simulation of a TiN film on M2 substrate considering energy of 1.1 MeV, 0 of -60° , composition of the Ti film 50% and N 50%, of the intermediate Ti film and composition of the M2 substrate, as described in Item 4.1.2.

The simulated spectrum of the TiN film deposited on tool support D6 shows a simulated spectrum (Figure 4.24) very similar to that of support M2 (Figure 4.23). The start of the detection of each element in the simulated RBS spectrum is indicated by an arrow.

These results show that it is possible to plan the deposition of films deposited on chromium and iron-based substrates and thus be able, using the new energy and 0 values presented in this work, to carry out RBS measurements in order to identify the elements relating to the chemical composition of the films with greater precision. This study is completely original and will be very useful for better control in the manufacture of functional films proposed in this thesis.

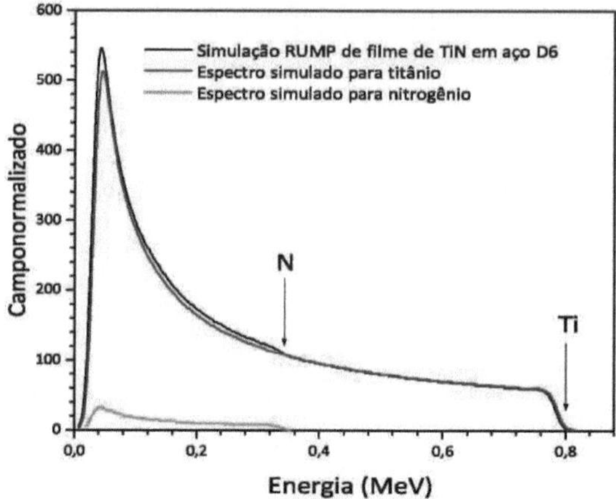

FIGURE4.24. Simulated spectra, considering the new energy values and 0, of a TiN film deposited on a D6 substrate, 50% Ti and 50% N film composition, intermediate Ti film and D6 substrate composition, as described in Item 4.1.2.

4.2.1.6. 4-Point Bending Test Curves of M2 and D6 Bodies with Deposited Titanium Nitride Films

The substrates of M2 and D6, with polished surfaces and titanium nitride films deposited at temperatures of 220 and 450° C, were subjected to the 4-point bending test. The
Photomicrographs of the samples, prepared according to the procedures adopted in Item 3.2.4.10, are shown in Figures 4.25 to 4.28. Despite an adequate search, no standard or reference was found in the literature that established the parameters for making specimens and/or conditions for bending tests for substrates with deposited films. It was therefore necessary to adapt the ASTM E855 - 90 standard[130] .

The SEM images of the surfaces of these films after the bending test are shown in Figures 4.25 to 4.28. The titanium nitride films deposited on the M2 and D6 steels at temperatures of 220 and 450° C show the presence of microcracks, since the deformation of this ceramic film is less than that of the metal substrate. These microcracks are unidirectional and orthogonal to the length of the sample and occur in greater numbers in the films deposited at 450° C (Figures 4.26 and 4.28). These cracks seem to start in pores, which is consistent with the greater fragility of the film. Clusters of varying sizes can be seen on the surfaces of all the samples, with the films deposited at the highest temperature showing a greater number and variation in the sizes of these surface defects. This is related to the higher deposition rate of the TiN films at 450° C .

In the TiN films deposited at a temperature of 220° C on apo M2, the presence of microcracks can be seen in the region close to the charge application lines, but in small quantities (Figures 4.25a and b). In the center of the sample (the region between the load application lines) no microcracks were detected. This may be due to the homogeneous distribution of stresses in this region of the specimen during the test (Figures 4.25c and d). Some regions of localized detachment are observed in the film with chaotic distributions and locations that are not associated with the regions of mechanical stress concentration. This effect seems to be associated with the detachment of the clusters.

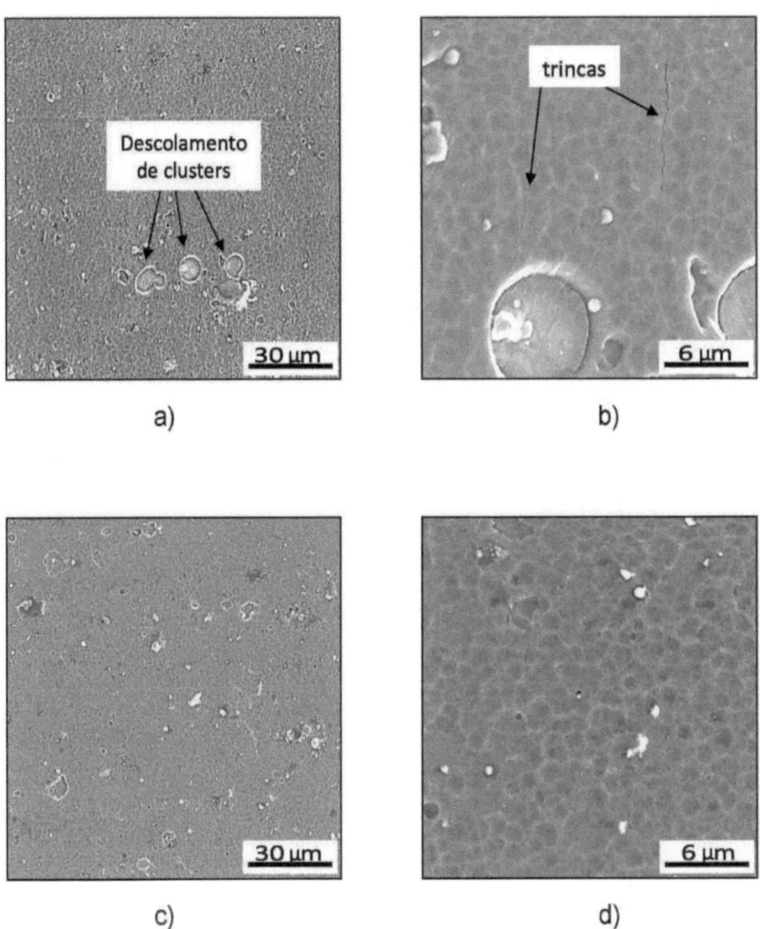

a)

b)

c)

d)

FIGURE 4 25 SEM photomicrographs of the polished surface of M2 steel with TiN film deposited at 220° C subjected to the 4-point bending test: a) and b) region where the load was applied to the TiN surface; c) and d) center of the sample on the TiN surface.
The surfaces of one of the M2 steel samples with titanium nitride film deposited at 450° C, after being subjected to the four-point bending test, are shown in Figure 4.26.

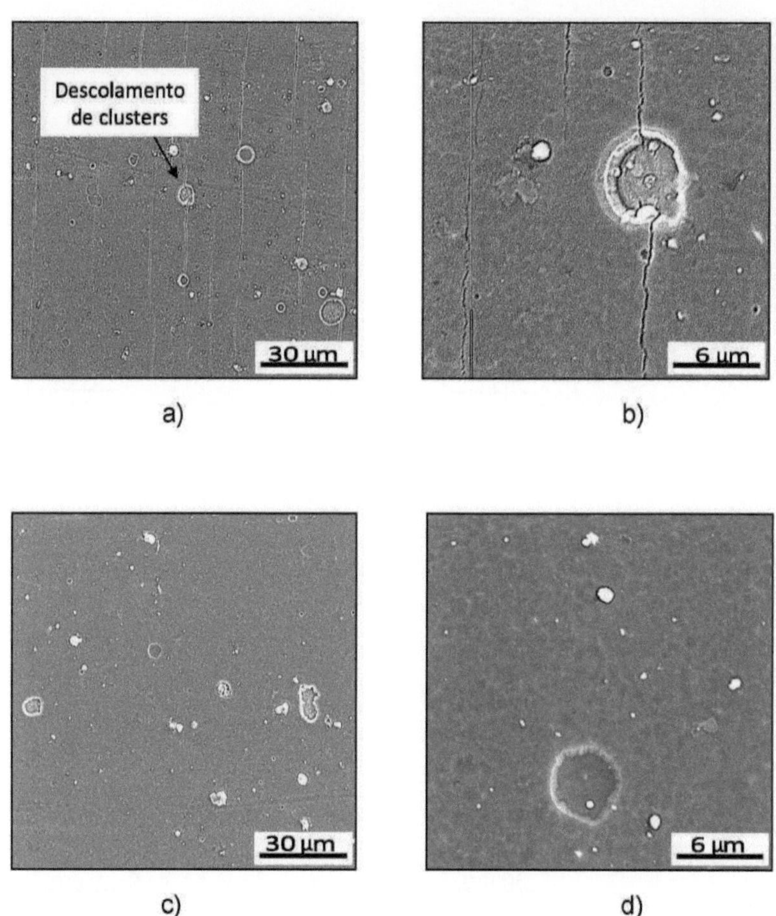

FIGURE 4 .26 Photomicrographs of the surface of M2 steel with TiN film deposited at a temperature of 450° C and subjected to the 4-point bending test: a) and b) region where the load was applied to the TiN surface; c) and d) center of the sample on the TiN surface.

This TiN film also shows the formation of unidirectional microcracks orthogonal to the length of the substrate, as in the previous sample, but in greater concentration located in the region close to the load application lines (Figures 4.26a and b). Defects can also be seen in the film which are associated with the detachment of clusters. It appears that microcracks begin in these defects. In the center of the sample (the region between the charge application lines), no microcracks are observed (Figures 4.26c and d).

The surfaces of one of the D6 steel samples with titanium nitride films deposited at 220° C, after the four-point bending test, are shown in Figure 4.27.

This TiN film also shows the formation of unidirectional microcracks orthogonal to the length of the substrate, as in the previous samples, with a small concentration located in the region close to the load application lines (Figures 4.27a and b). These microcracks seem to start in the regions where clusters are found and in the defects arising from the detachment of clusters. This is quite consistent given the greater fragility of this film. In the center of the sample (the region between the charge application lines) no microcracks are observed (Figures 4.27c and d), but localized detachment of the film by clusters is observed.

Regions of the surface of one of the apo D6 samples with a titanium nitride film deposited at 450° C, after the four-point bending test, are shown in Figure 4.28.

As can be seen in Figures 4.28a and b, this TiN film also shows the formation of unidirectional microcracks orthogonal to the length of the substrate, as in the previous samples, with a large concentration located in the region close to the load application lines. Microcracks also seem to form in the cluster regions, in greater quantity and width when compared to the previous tests. In the center of the sample (the region between the lines where the fillers were applied), no microcracks formed (Figures 4.28c and d) and localized detachments of the film were observed.

FIGURE 4 .27 Photomicrographs of the surface of the TiN film deposited at 220° C on D6 steel, obtained by SEM, subjected to the 4-point bending test: a) and b) region of the load application on the TiN surface; c) and d) center of the sample on the TiN surface.

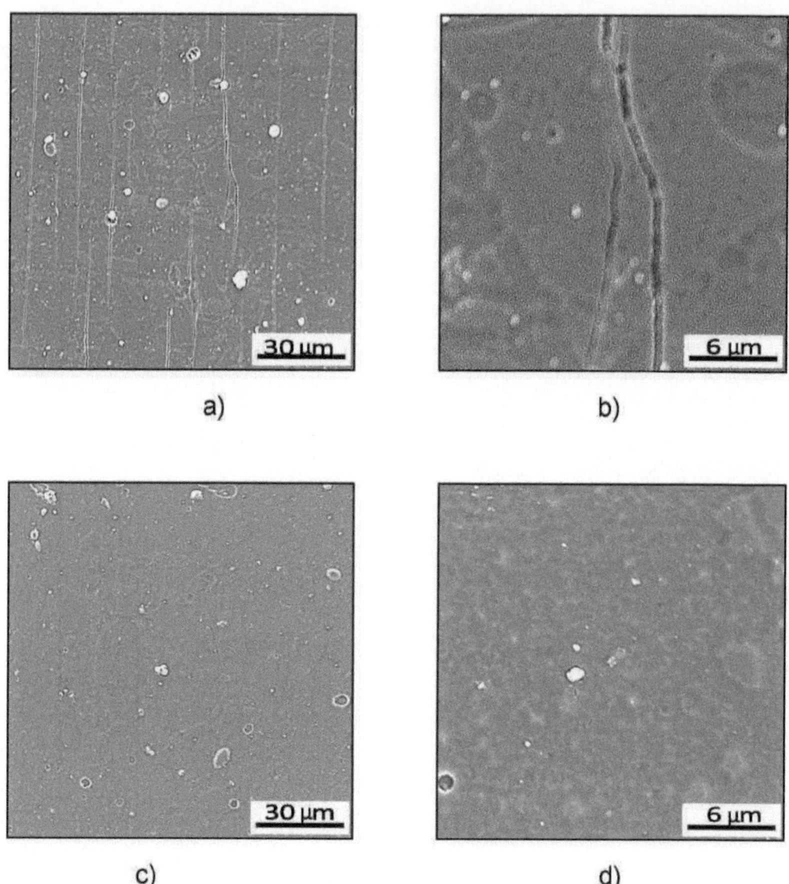

FIGURE 4 .28 SEM photomicrographs of the polished surface of the TiN film deposited at a temperature of 450° C on D6 steel and subjected to the 4-point bending test: a) and b) region where the load was applied to the TiN surface; c) and d) center of the sample on the TiN surface.

Stress versus strain curves were obtained by 4-point bending for the substrates of the apo M2 and apo D6 tools with deposited TiN films. The stress versus strain curves for the substrates of apo M2 and apo D6 with TiN films deposited on them were also obtained.

TiN have similar mechanical behavior in relation to the film deposition temperatures. The substrates with TiN films deposited at 220° C showed higher bending stress values (from the yield stress of the assembly) than those with films deposited at 450° C (Figures 4.29 and 4.30). This result may be associated with the greater thickness of the films deposited at 450° C and/or changes in the substrate microstructures of the M2 and D6 samples.

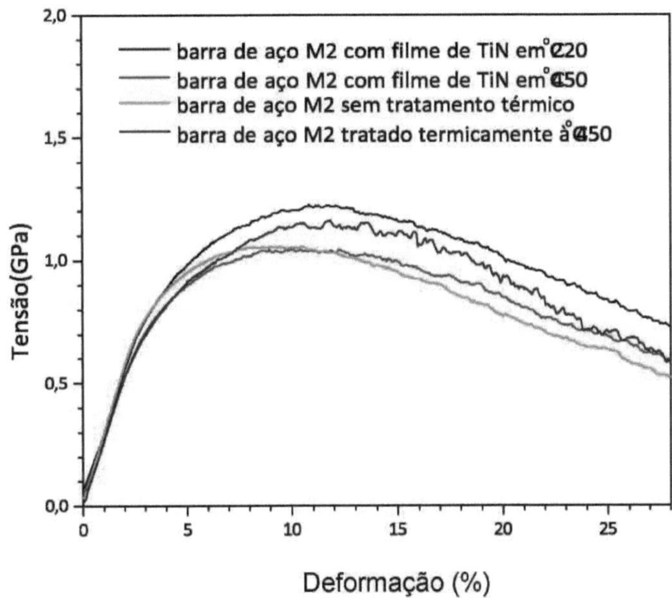

FIGURE 4 .29 Stress versus strain curves obtained from the 4-point bending tests for apo M2 substrates with TiN films deposited at 220 and 450 C.°

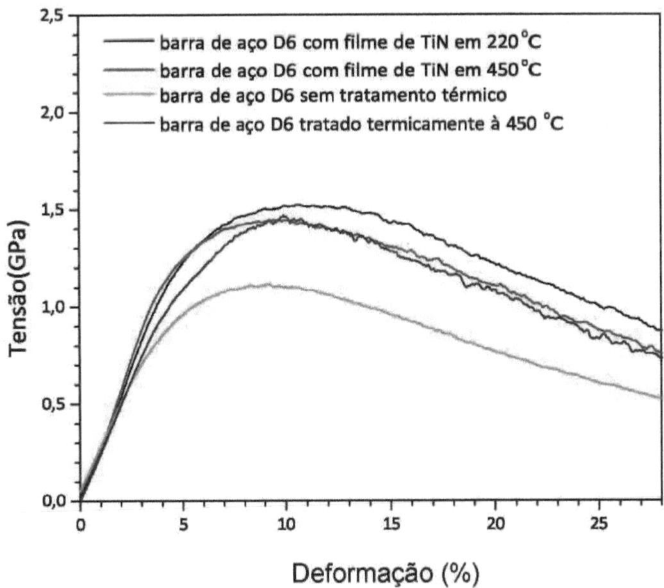

FIGURE 4 .30 4-point bending test curves for D6 steel samples with TiN films deposited at 220 and 450 C.° The curve obtained for the titanium nitride film grown at a temperature of 220° C on the surface of M2 steel, throughout the 4-point bending test, does not show the same behavior as the substrate. The smaller thickness obtained at this temperature may not have been sufficient for this type of measurement.

On the other hand, the curve of the titanium nitride film deposited at a temperature of 450° C on M2 steel, for deformation values of around 23.7%, tends to show the behavior of the M2 steel substrate material. This behavior can be related to

with rupture and/or detachment of the titanium nitride film. The stresses associated with these changes in the behavior of the curves are approximately 0.74 GPa, for M2 steel with a titanium nitride film deposited according to the conditions described in Table 3.1.

The curve obtained for the titanium nitride films grown at a temperature of 220° C on the surface of D6 steel also does not show the same behavior as the substrate throughout the 4-point bending test.

For deformation values of around 9.14 %, the curve of the titanium nitride films deposited at a temperature of 450° C on the surface of D6 steel tends to show the same behavior as the material on the D6 steel substrate. This behavior can be related to the rupture and/or detachment of the titanium nitride film. The stresses associated with these changes in the behavior of the curves for D6 steel with titanium nitride film, deposited under the conditions described in Table 3.1, are approximately 1.44 GPa.

According to these measurements, it can be seen that the titanium nitride film deposited at a temperature of 450° C on the M2 steel substrate resisted rupture for longer than the same film deposited on the D6 steel substrate. This is possible because the titanium nitride film deposited on M2 steel at this temperature took longer for the layer closest to the surface to peel off (Item 4.2.1.2).

In these analyses, it was observed that, with the higher deposition temperature of this titanium nitride film, it was possible to verify the exact moment of film detachment during the test. This may be associated with the larger interface region of the titanium nitride film promoted by the higher temperature of the M2 and D6 substrates used during the deposition process. Therefore, the use of this technique will only be interesting for film/substrate systems that have a satisfactory interface region. On the other hand, the greater thickness produced on the M2 and D6 steel surfaces at temperatures of 450° C also seems to have contributed to the good reading of the moment of film detachment by this technique. However, the defects observed on the surface of the TiN film/D6 steel substrate system obtained at 450° C (Item 4.2.1.2.) caused this film to break before the same film system deposited on M2 steel. In this case, it is necessary to better control the deposition parameters of these films in order to strike a balance between the formation of a larger interface region and the quality of these films, avoiding the formation of defects.

This technique for measuring the adhesion of films to substrates is very promising. However, further studies are needed to determine the breakdown voltage and the conditions for the desired interfaces in film/substrate systems.

4.2.I.7. Partial conclusions on TiN films

The results of the X-ray diffractograms show that the deposition temperature contributed strongly to the nucleation and growth of the titanium nitride films deposited on the surfaces of M2 and D6. On the other hand, the substrates do not seem to have contributed to the nucleation and growth of these films.

SEM images show that titanium nitride films deposited at 220° C (lower temperatures) are more homogeneous and have denser structures.

The AFM results show that the titanium nitride films deposited at 220° C have a much more homogeneous microstructure with nanograins of similar size and shape. These films deposited at this temperature have a nanostructure.

The electron backscattering results showed that an interface with fewer defects was formed for the titanium nitride films deposited at the lower temperature.

The stoichiometry of these films, presented by RBS, was very good in all the films deposited. However, it was also possible to identify a concentration of contaminants, such as carbon. To eliminate this contaminant, it is necessary to use a vacuum system that works at higher vacuum pressures and in a clean way, without the use of oil.

In this work, we present a new technique for measuring films grown on substrates with elements of greater atomic weight than the components of the deposited films, and in a more precise way. This type of study is completely original and will be useful in the development of any work in the area of films grown on these types of substrates, as well as for future work on the growth of functional films, as presented in this thesis.

The films deposited at 450° C in apo M2 had a higher resistance to rupture when compared to the same film deposited at the same temperature in apo D6, showing that this temperature is also very good for deposits in apo M2.

The results of the 4-point bending tests indicated excellent adhesion of the titanium nitride films deposited at 450° C, with tensile strength and/or detachment values of 3.12 and 5.96 GPa for the films deposited at M2 and

D6, respectively. The small thickness obtained in the depositions at 220° C hindered the measurement by this technique.

Table 4.6 shows the relationship between the type of substrate and the deposition temperature of this film.

TABLE 4.6. Degrees of characteristics of the TiN films deposited.

Film type	Substrate	Temperature (° C)	Movie quality	Interface quality	Thickness (nm)	Adherence to the interface
TiN	M2	220	good	good	= 1200	good
	D6	220	good	good	= 2100	good
	M2	450	regular	regular	= 2400	good
	D6	450	terrible	regular	= 3000	good

It was expected that these films would be deposited with the same thickness on all the samples. However, according to the results shown in Table 4.6, the type of substrate greatly influenced the thickness of the film deposited, being greater on D6 steel substrates, regardless of temperature. As expected, the temperature influenced the quality of the film and the interface formed. As a result, only films grown at a lower temperature were able to show better growth results overall.

The intermediate titanium film helped to increase the adhesion of these TiN films to the substrates, as it would have helped to form the interface region, promoting the adhesion of these films. However, although the TiN films grown at a temperature of 450° C showed detachment near the surface, their interface remained without any signs of detachment. Therefore, even with a very small thickness, the function of the intermediate titanium film is of great importance in this type of ceramic/metal deposition. The detachment of the region close to the surface of the TiN films deposited on D6 steel at a temperature of 450° C may be associated with the greater thickness obtained during deposition.

In the bending tests, the films grown at a lower temperature did not break during the test. The smaller thickness may have hindered this type of measurement. However, even the film grown at a temperature of 450° C in apo D6, which was the worst case, broke after a stress of 5.96 GPa, proving once again the importance of the intermediate film used in this deposition and good formation of the interface region.

4.2.2. Aluminum Nitride Films

4.2.2.1. Characterized by X-ray diffraction

Conventional X-ray diffraction analysis was carried out on the aluminum nitride films deposited on the surfaces of the M2 and D6 plates under the conditions described in Table 3.1.

The X-ray diffractograms of the apo M2 substrates show the presence of the iron ferrite phase and the M6C phase (Figure 4.31).

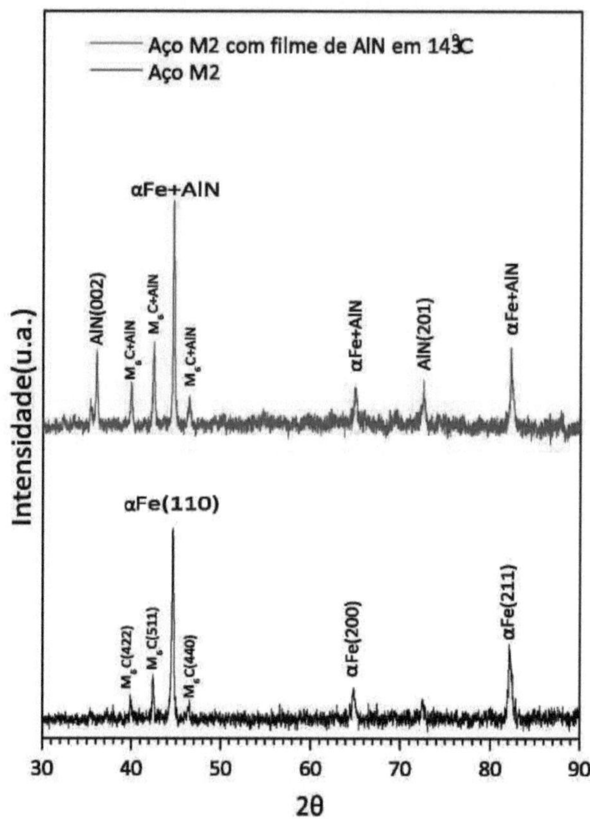

FIGURE 4 .31 X-ray diffractograms of the apo M2 substrate and
the aluminum nitride films deposited on apo M2.

For the samples with deposited films, the X-ray diffractograms also show the presence of AlN, in addition to the phases present in the apo substrate. As can be seen in Figure 4.31, the apo M2 samples with AlN films deposited under the conditions described in Table 3.1 show the presence of AlN(002) and AlN(201), indicating the growth of AlN crystals in two crystalline planes preferentially.

Figure 4.32 shows the X-ray diffractograms of the apo D6 substrates and the AlN film deposited on apo D6.

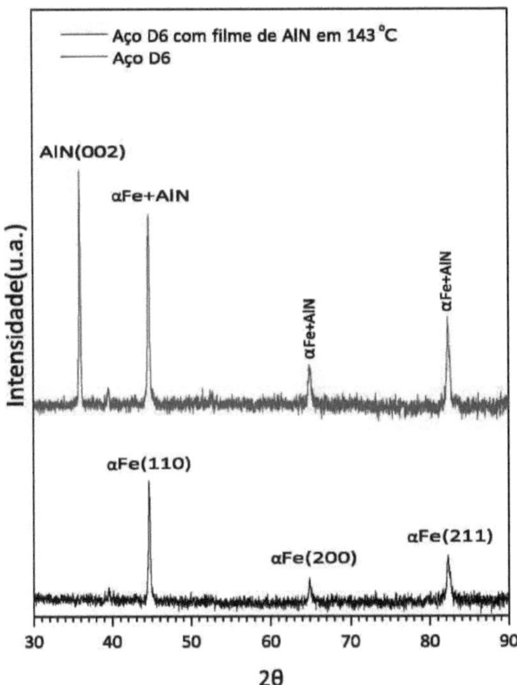

FIGURE 4 .32 X-ray diffractograms of the aluminum nitride film deposited on the surface of apo D6 and the apo D6 substrate.

As can be seen, the apo D6 samples with deposited AlN films show the AlN(002) peak, indicating the growth of AlN crystals in a preferential plane.

4.2.2.2. Characterized by SEM and EDX

The intermediate aluminum films and aluminum nitride films were deposited on the M2 and D6 substrates at 143° C and under the conditions described in Table 3.1.

The aluminum nitride films deposited on the apo M2 substrates show the presence of shallow scratches in orthogonal directions (Figures 4.33a, b and c). The EDX spectrum obtained for this surface is shown in Figure 4.33d, showing the presence of the chemical elements that make up the film and the substrate. This indicates that there was diffusion of iron and chromium into the AlN, which may have influenced the surface roughness of the film.

The aluminum nitride films deposited on the apo D6 substrates show deep scratches all over their surface in directions orthogonal to each other, but no clusters (Figures 4.34a, b and c). The spectral curve obtained by EDX for this surface, showing the presence of the component elements of the film and the substrate, is shown in Figure 4.34d.

Subsequent topographical observations, using SEM, of the surfaces of the substrates with AlN films deposited on the surfaces of the M2 and D6 samples (Table 3.1) showed the film peeling off, as can be seen in Figure 4.35.

After a short time, they in turn became completely detached from the surface of the substrates. In this case, it was not possible to carry out AFM, XPS, RBS and 4-point bending analyses of these surfaces.

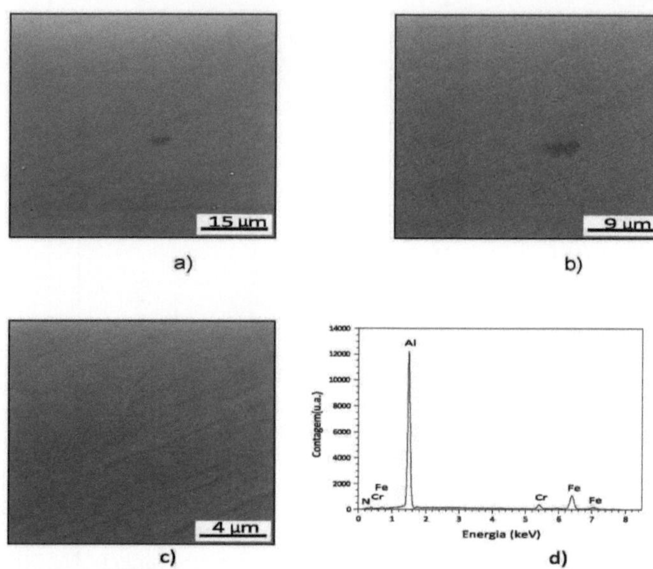

FIGURE 4 .33 SEM photomicrographs of the AlN film deposited on the surface of M2 steel, a) 2000x, b) 3500x and c) 7500x and d) spectral curve obtained by EDX of the surface of this film.

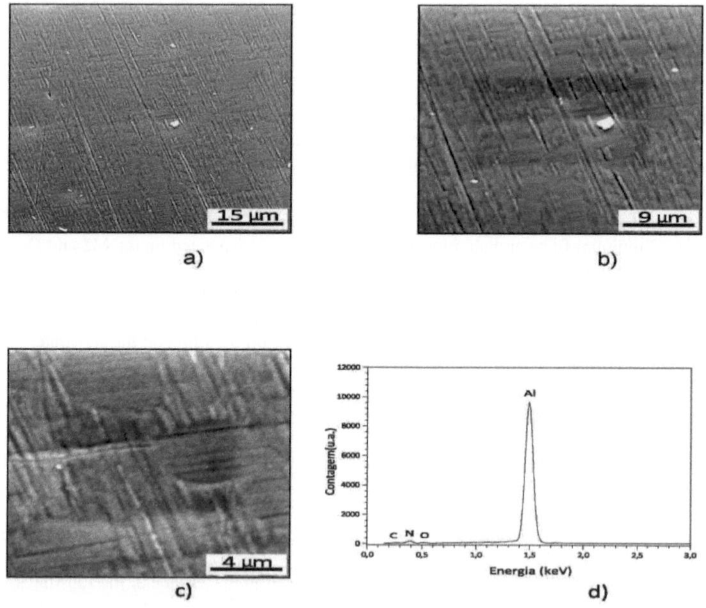

FIGURE 4 .34 Photomicrographs for the AlN film deposited on the surface of apo D6 obtained by SEM, a) 2000x, b) 3500x and c) 7500x and d) Spectrum of the surface of this film obtained by EDX.

a)   b)

FIGURE 4 .35 SEM photomicrographs of the AlN film deposited under the conditions in Table 3.1: a) on apo M2 and b) on apo D6.

4.2.2.3. Partial Conclusions on Aluminum Nitride Films

Table 4.7 shows the influence of adhesion on the type of substrate used and the deposition temperature of the film.

TABLE 4.7. Degrees of characteristics of AlN films deposited.

Type of film	Substrate	Temperature (° C)	Movie quality	Thickness (nm)	Quality of adhesion
AlN	M2	143	good	500	terrible[*]
	D6	143	good	500	terrible[*]

(*) The films peeled off immediately after deposition.

The aluminum nitride films deposited on M2 and D6 substrates, using an intermediate layer of aluminum, did not achieve satisfactory adhesion. The film peeled off completely from the surface of the substrates after a short time. This detachment may be associated with the lack of cleaning of the substrates by sputtering, which in this case was not available in the deposition equipment used (Item 3.2.3.2.). Another factor may be related to the great avidity of the aluminum to react with oxygen, which ended up competing with the nitrogen present in the chamber. In the next sections, we will study the deposition of this type of film in a functional way, creating regions in which the properties of the film and the substrate vary gradually, in order to create zones to absorb the tensions generated by the deposited film.

4.2.3. Characterization of TiN and AlN Functional Films

4.2.3.1. Characterization of Functional TiN and AlN Films by X-Ray Diffraction

FIGURA 1.1. Titanium Nitride Functional Films

Conventional X-ray diffraction analysis was carried out on the functional titanium nitride films deposited on the surfaces of apo M2 and D6. Figure 4.36 shows the X-ray diffractograms of the apo M2 substrates and the TiN films deposited on
apo M2, under the conditions of pressure 2-2.3 mTorr, power 200 W and substrate temperature 177° C (Condition 1 - Table 3.2). These films were grown by varying the argon and nitrogen concentrations of 50 /

40 / 35 / 30 / 25 and 0 / 10 / 15 / 20 / 25 sccm, respectively.

The X-ray diffractograms of the M2 apo substrates show the presence of the ferrite phase and the M6C phase. For the film-deposited samples, in addition to the crystalline phases present in the apo substrate, the X-ray diffractograms also show the presence of TiN. As can be seen, the apo M2 samples with TiN films, deposited according to the conditions specified in Table 3.2, show the TiN(200) and Ti2N(202) peaks, indicating the nucleation and growth of TiN crystals in two preferential crystalline directions. The main characteristic of the Ti-N system is the existence of a large number of chemical compounds. The main one is TixNy, which dissolves 20 to 55 % of nitrogen atoms and which explains, despite being a metastable crystalline phase, the presence of the Ti2N(202) compound in this deposited film[125,136].

Figure 4.37 shows the X-ray diffractograms of the apo D6 substrates and the TiN functional films deposited on apo D6 in Condition 1.

The X-ray diffractograms of the apo D6 substrates show only the presence of ferrite on the substrate. For the samples with deposited films, in addition to the phases present in the apo substrate, the X-ray diffractograms also show the presence of TiN. As can be seen, the apo D6 samples with TiN films deposited according to the conditions specified in Table 3.2 show the TiN(200) peak, indicating the nucleation and growth of TiN crystals in a preferential direction.

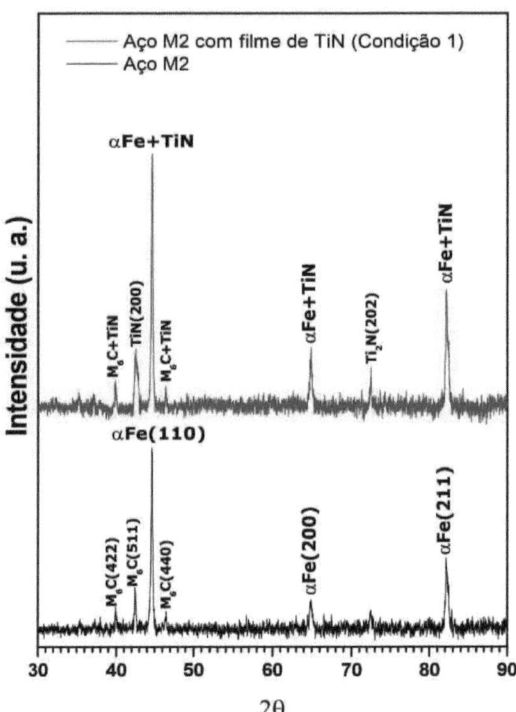

FIGURE 4 .36 X-ray diffractograms of the apo M2 substrate and the titanium nitride films deposited on apo M2 in Condition 1.

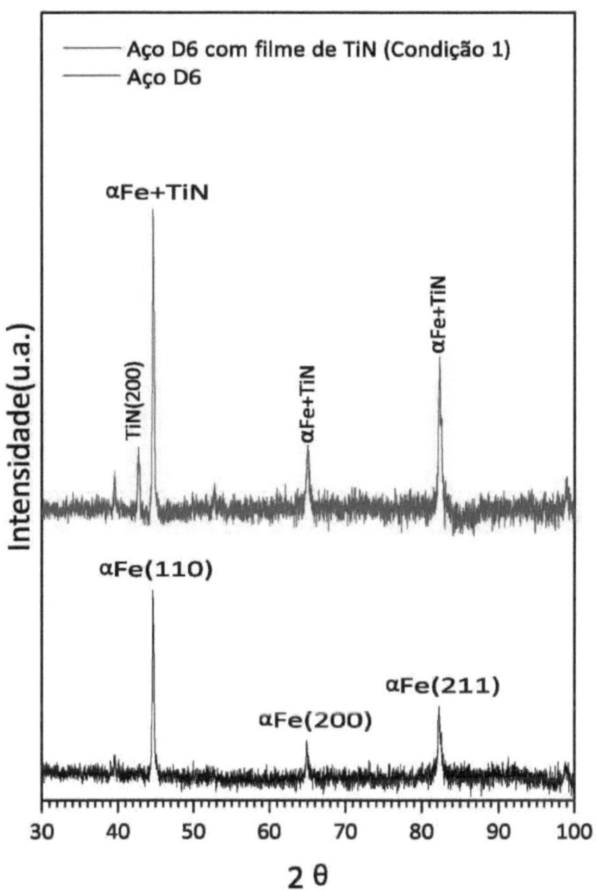

FIGURE 4 .37 X-ray diffraction curves of the TiN films deposited on apo D6 (Condition 1) and the apo D6 substrate.

4.2.3.1.2 Functional Aluminum Nitride Films

Figure 4.38 shows the X-ray diffractograms of the apo M2 substrates and the functional AlN films deposited on apo M2 under conditions 2 to 4. The conditions used for these films are: pressure of 1.5 mTorr, power of 200 W and substrate temperature of 114° C for Conditions 2 to 3, and 135° C for Condition 4. The films were designed by varying the argon and nitrogen concentrations of 50 / 25 / 30 / 35 / 40 and 0 / 25 / 20 / 15 / 10 sccm, respectively, for Condition 2, and 50 / 40 / 35 / 30 / 25 and 0 / 10 / 15 / 20 / 25 sccm, respectively, for Condition 3, and 50 / 0 / 0 / 0 and 50 / 10 / 15 / 20 / 25 sccm, respectively, for Condition 4. The thickness of all the functional films is estimated at 250 nm, according to the procedure already described in Item 3.2.3.2.

The X-ray diffractograms of the M2 steel substrates show the presence of ferrite and the crystalline phase M6C. For the samples with deposited films, in addition to the phases present in the steel substrate, the X-ray diffractograms also show the presence of AlN. As a result of the analysis of these diffraction peaks, the M2 steel samples with AlN films, deposited according to the process parameters specified in Table 3.2, show peaks relating to the AlN(002) and AlN(201) crystal planes, indicating the nucleation and growth of AlN crystals in two preferential crystal directions.

Figure 4.39 shows the X-ray diffractograms of the apo D6 substrates and the AlN functional films deposited on apo D6 in Conditions 2 to 4.

The X-ray diffractograms of the apo D6 substrates show the presence of the ferrite phase (iron a). For the samples with the films deposited, in addition to the phases present in the apo substrate, the X-ray diffractograms also show the presence of AlN. As can be seen, the apo D6 samples with AlN films deposited according to the conditions specified in Table 3.2 show the AlN(002) peak, indicating the nucleation and growth of AlN crystals in a preferential crystalline direction. However, for the AlN films deposited in Condition 4 (without argon in the plasma), the AlN(101) peak was formed. This shows that, in the absence of argon gas in the plasma, during the formation of the functional AlN film, there is a tendency for other planes to grow preferentially. As there is no more argon reaching the aluminum target, it in turn starts a saturation process forming AlN compounds on its surface, even before they are released onto the substrate surface.

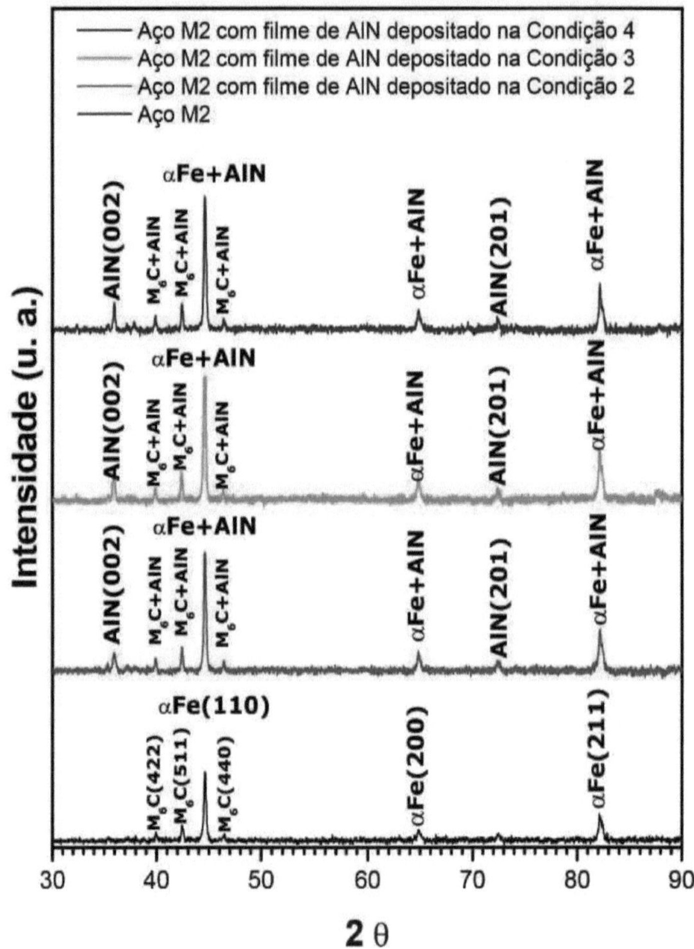

FIGURE 4 .38 X-ray diffractograms of the AlN films deposited on apo M2 under conditions 2 to 4 and of the apo M2 substrate.

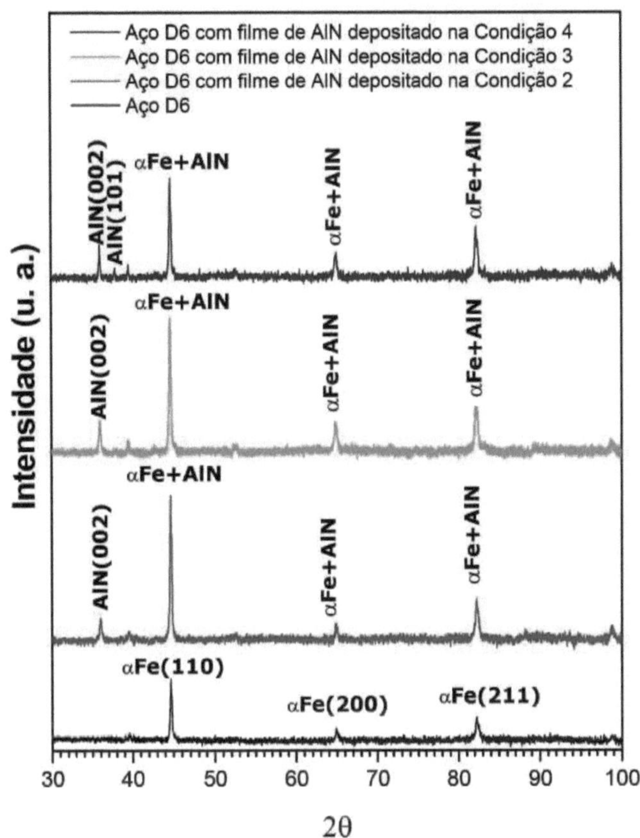

FIGURE 4 .39 X-ray diffraction curves of the apo D6 substrate and the AlN films deposited on apo D6 in Conditions 2 to 4.

4.2.3.2. Characterization of Functional Films by SEM and EDX

4.2.3.2.1. Functional Titanium Nitride Films

The titanium nitride films deposited on the M2 steel substrates, using the sequence of argon and nitrogen gas concentrations in the plasma, in Condition 1, according to the conditions specified in Table 3.2 (temperature of 177° C, deposition time of 1h and estimated thickness (Item 3.2.3.2) of 250 nm), show a surface with many scratches and pores, with homogeneous pore size distribution (Figures 4.40a, b, and c).

The lighter grains and black dots, according to the EDX analysis already presented in Section 4.1.1, are composed of a greater amount of molybdenum, tungsten and vanadium than the matrix (darker region). These chemical elements are normal components of M2 high-speed steel (according to Items 3.1 and 4.1.2) and, as the film is very thin, these grains can be observed by SEM. The spectral curve obtained by EDX for this surface is shown in Figure 4.40d, proving the presence of the chemical elements that make up the film and the substrate.

The titanium nitride films deposited on the D6 steel substrates, using the sequence of argon and nitrogen gas concentrations in the plasma, in Condition 1, according to the conditions specified in Table 3.2 (temperature of 177° C, deposition time of 1h and estimated thickness (Item 3.2.3.2) of 250 nm), show a surface with shallow scratches and pores (Figure 4.41a, b, and c). However, there are also some darker regions on the surface of the film with a heterogeneous size distribution.

EDX analysis shows that these darker regions have a higher concentration of chromium, as already discussed in Item 4.1.1. The EDX spectrum obtained for this surface is shown in Figure 4.41d, showing the presence of the chemical elements that make up the film and the substrate.

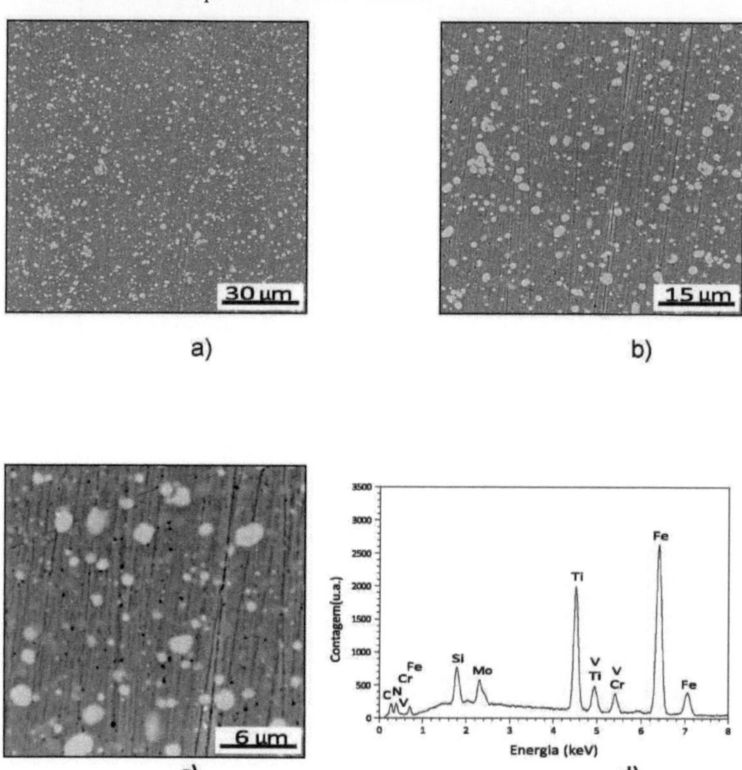

FIGURE 4 .40 SEM photomicrographs of the TiN functional film deposited on the surface of apo M2 in Condition 1 (Table 3.2), a) 1000x, b) 2000x and c) 5000x and d) spectral curve obtained by EDX of the surface of this film.

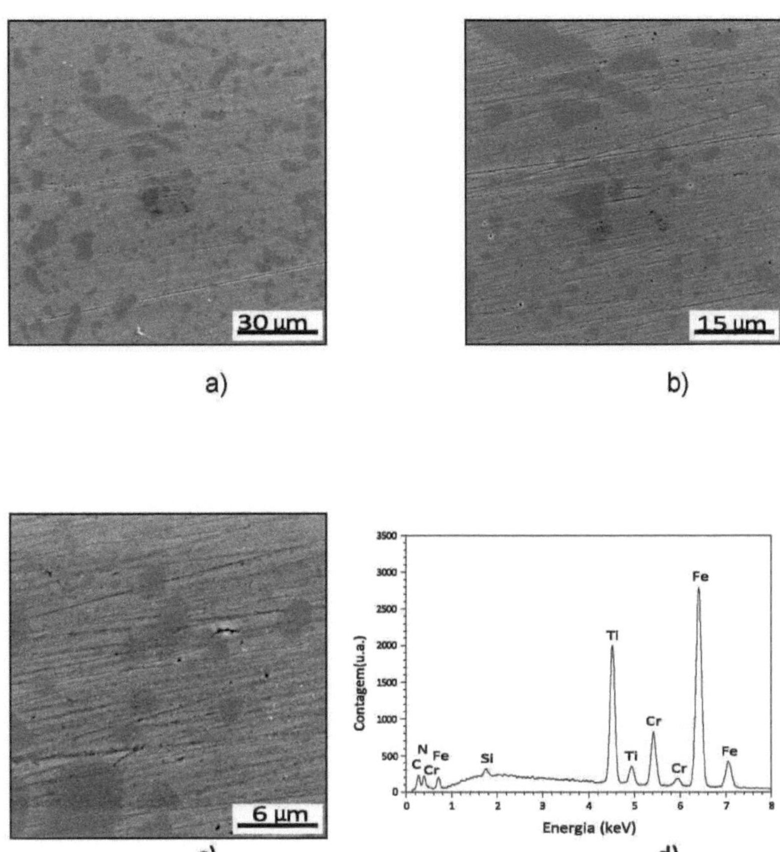

FIGURE 4 .41 Photomicrographs obtained in SEM for the TiN functional film deposited, in Condition 1 (Table 3.2), on the surface of D6 steel a) 1000x, b) 2000x and c) 5000x and d) EDX spectrum of the surface of this film.

4.2.3.2.2. Functional Aluminum Nitride Films

The aluminum nitride films deposited on the M2 steel substrates in Condition 2 (temperature of 114° C, deposition time of 1h and estimated thickness (Item 3.2.3.2) of 250 nm) according to the conditions specified in Table 3.2, show small pores, scratch-like defects and a fairly homogeneous distribution of grains, but with different sizes (Figures 4.42a, b, and c).

Spectral curves obtained by EDX for this surface are shown in Figure 4.42d, proving the presence of the chemical elements that make up the film and the substrate.

The aluminum nitride films deposited on the D6 steel substrates in Condition 2 (temperature of 114° C, deposition time of 1h and estimated thickness of 250 nm) according to the conditions specified in Table 3.2, have a more homogeneous film surface (Figure 4.43a, b, and c) when compared to the film deposited on the M2 steel. However, some regions show defects in the form of scratches, up to around 20 pm in length, and some pores of varying shapes. However, there is less evidence of smaller scratches (Figure 4.43c) when compared to the surface of the film deposited on the M2 steel substrate, which also shows the formation of scratches. The formation of these scratches may be associated with the increasing concentration of argon gas in the chamber which, through a sputtering process, ends up causing these defects to form on the surface of the substrates.

The EDX analysis for this surface is shown in Figure 4.43d, proving the presence of the chemical elements that make up the film and the substrate.

The aluminum nitride films deposited on the M2 steel substrates, reversing the sequence of the amount of argon and nitrogen gases in the plasma according to Condition 3 in Table 3.2 (temperature of 114° C, deposition time of 1h and estimated thickness of 250 nm), have a surface very similar to that obtained using Condition 2, but with fewer pores (Figure 4.44a, b, and c).

The spectral curve obtained by EDX for this surface, showing the presence of the chemical elements that make up the film and the substrate, is shown in Figure 4.44d.

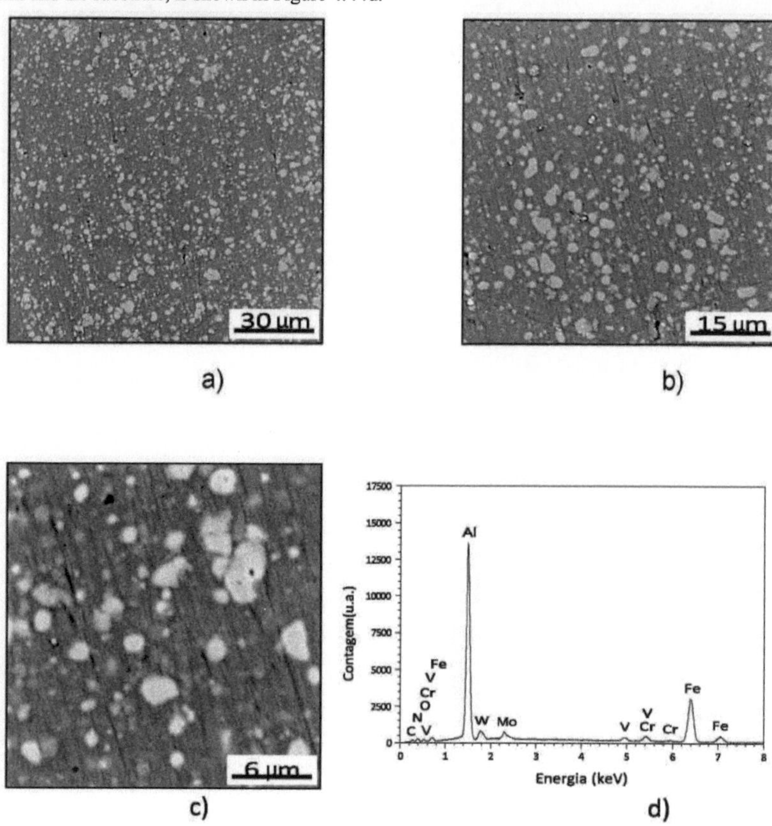

FIGURE 4 .42 SEM photomicrographs of the functional AlN film deposited on the surface of M2 steel in Condition 2 (Table 3.2): a) 1000x, b) 2000x and c) 5000x and d) spectral curve obtained by EDX of the surface of this film.

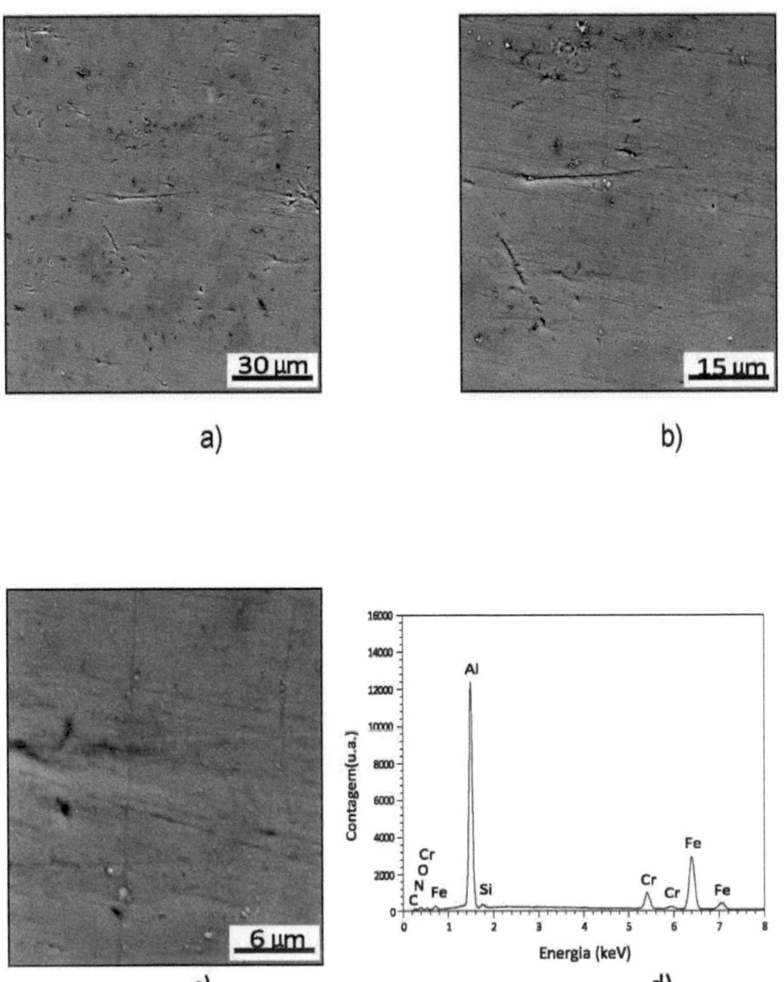

FIGURE 4 .43 Photomicrographs obtained by SEM of the surface of the AlN film deposited in Condition 2 (Table 3.2) on D6 steel: a) 1000x, b) 2000x and c) 5000x and d) spectral curve of the surface of this film obtained by EDX.

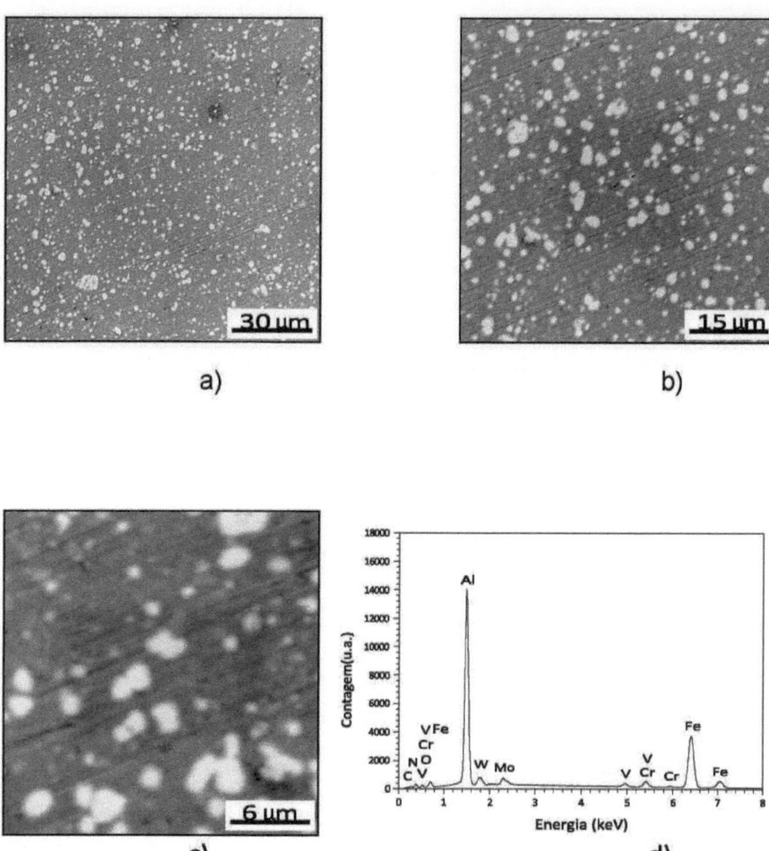

FIGURE 4 .44 SEM photomicrographs of the functional AlN film deposited on the surface of M2 steel in Condition 3 (Table 3.2): a) 1000x, b) 2000x and c) 5000x and d) EDX spectrum of the surface of this film.

The aluminum nitride films deposited on the apo D6 substrates, also by reversing the concentration sequence of the argon and nitrogen gases in the plasma according to Condition 3 of Table 3.2 (temperature 114° C, deposition time 1h and estimated thickness 250 nm), show defects in the form of scratches (Figures 4.45a, b, and c). However, the surface of the film shows some darker regions with a wide distribution of sizes and heterogeneous shapes over the film matrix. EDX analysis showed that these regions have the same composition as the film/substrate system but with a higher concentration of chromium (according to Item 4.1.1). The spectral curve obtained by EDX for this   surface is shown in Figure 4.45d,   proving the presence of the chemical elements film and substrate components.

With the change from condition 2 to 3, increasing the amount of nitrogen in the plasma, there is a slight decrease in the amount of pores and defects in the form of deeper scratches. This may be associated with the decreasing concentration of argon gas inside the chamber as the nitrogen gas increases. This reduces the possibility of argon ions causing defects on the substrate surface by sputtering.

The aluminum nitride films deposited on the apo M2 substrates, maintaining the last nitrogen gas concentration sequence used, but without the use of argon gas in the plasma, according to Condition 4 of Table 3.2 (temperature of 135° C, deposition time of 1h and estimated thickness of 250 nm), they show few small pores, shallow scratch-like defects and an inhomogeneous grain size distribution (Figures 4.46a, b, and c). The presence of the chemical elements that make up the film and the substrate, as evidenced by the spectral curve obtained by EDX for this surface, is shown in Figure 4.46d.

The deposition of aluminum nitride films on D6 steel substrates, maintaining the last nitrogen gas concentration sequence used, but without the use of argon gas in the plasma, according to Condition 4 of Table 3.2 (temperature of 135° C, deposition time of 1 hour and estimated thickness of 250 nm), show defects in the form of shallow scratches and an inhomogeneous grain size distribution (Figure 4.47a, b, and c). However, the photomicrographs show the presence of some darker regions on the surface of the film with a heterogeneous distribution of sizes and shapes, similar to those shown in the films deposited in Condition 3 on the D6 steels (Figure 4.45). The EDX analyses also show a higher concentration of chromium in these regions, as discussed in Item 4.1.1. The spectral curve obtained by EDX for this surface is shown in Figure 4.47d, showing the presence of the chemical elements that make up the film and the substrate.

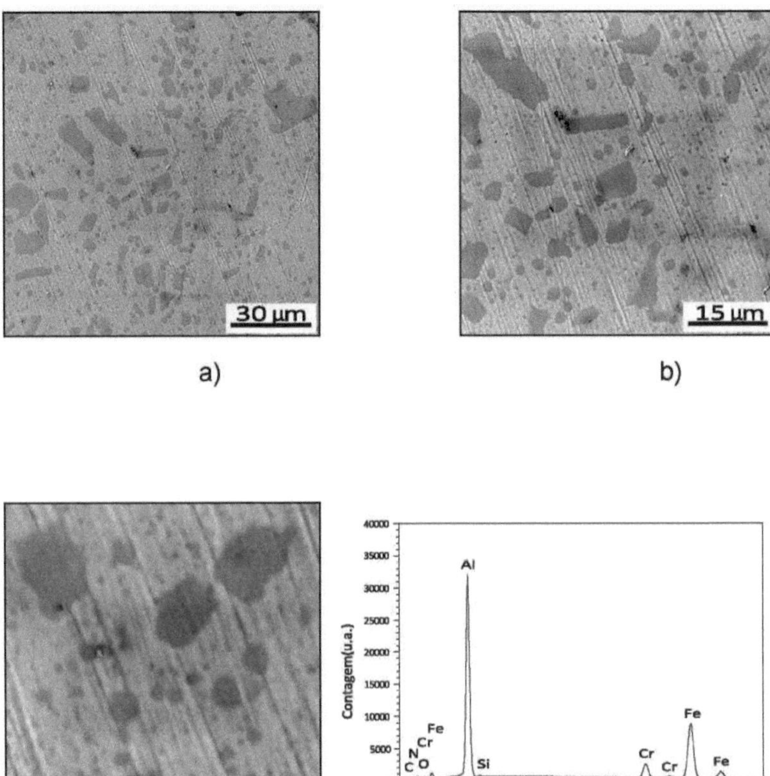

FIGURE 4 .45 Photomicrographs obtained in SEM for the AlN film deposited in Condition 3 (Table 3.2) on the surface of D6 steel: a) 1000x, b) 2000x and c) 5000x and d) EDX spectral curve of the surface of this film.

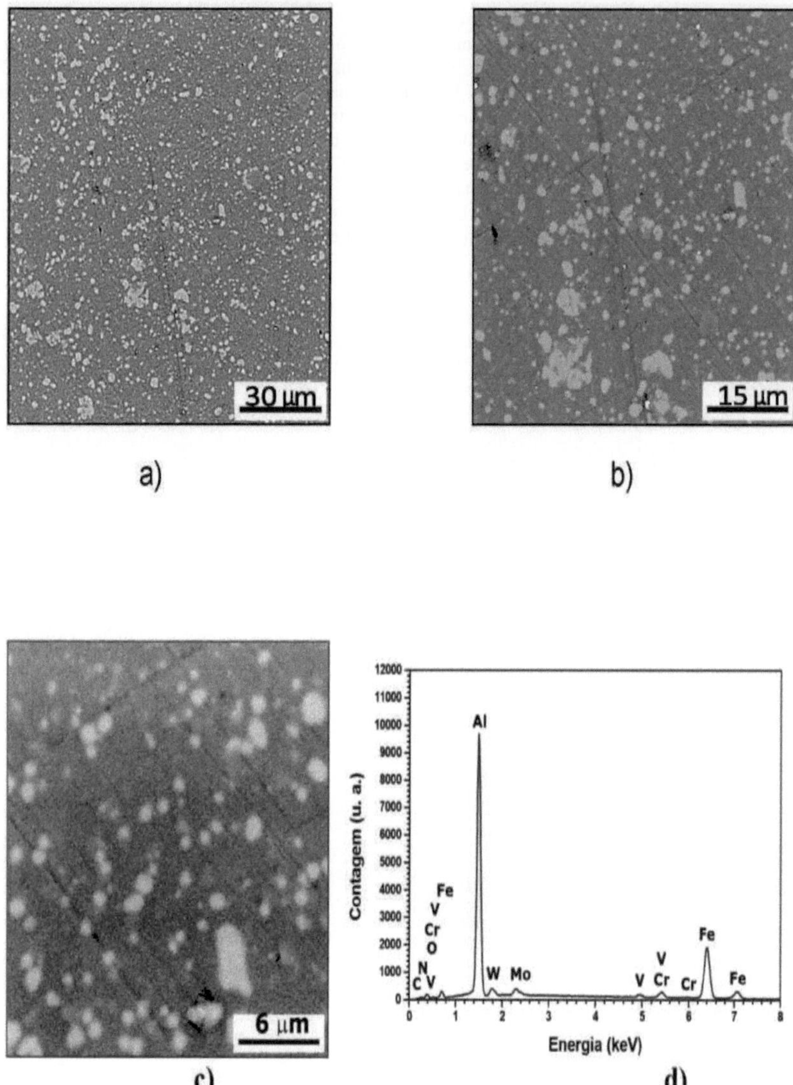

FIGURE 4 .46 SEM photomicrographs of the surface of the functional AlN film deposited on the M2 steel substrate in Condition 4 (Table 3.2): a) 1000x, b) 2000x and c) 5000x and d) spectral curve of the surface of this film obtained by EDX.

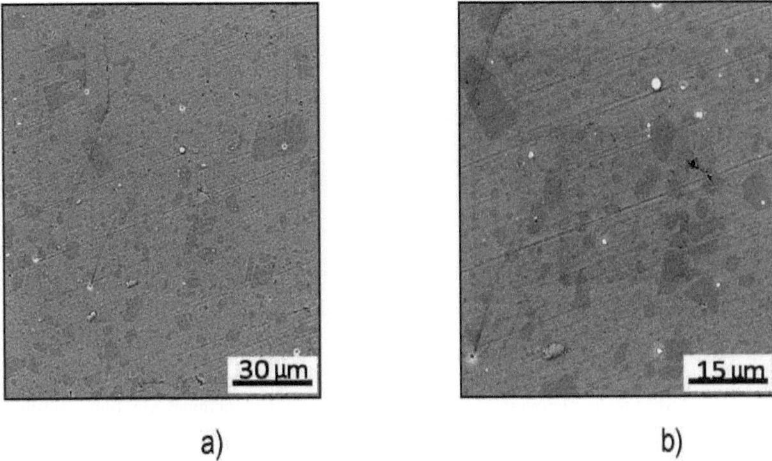

a)   b)

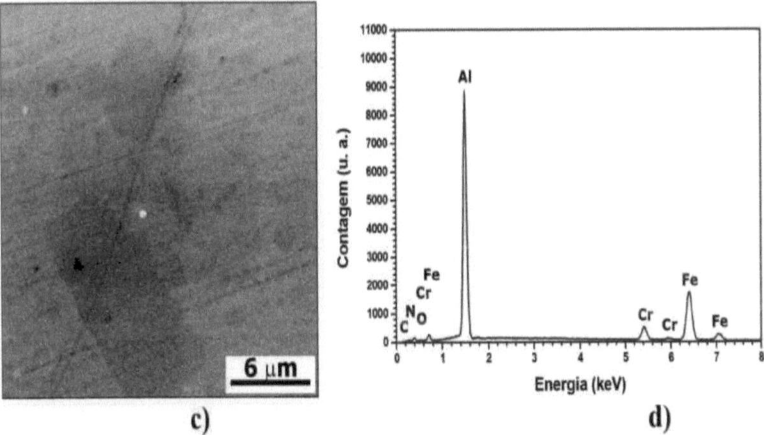

c)   d)

FIGURE 4 .47 SEM photomicrographs of the surface of the AlN functional film deposited in Condition 4 (Table 3.2) on the D6 steel substrate: a) 1000x, b) 2000x and c) 5000x and d) EDX spectral curve of the surface of this film.

Subsequent topographical observations of the surfaces of the functional AlN films deposited on the surfaces of M2 and D6 steels (Conditions 3 and 4 - Table 3.2), using the SEM technique, show that the film has peeled off in defined regions, as can be seen in Figure 4.48. The films deposited under conditions 3 and 4 have a similar topography. The functional AlN film deposited in Condition 2 detached completely from the steel surfaces. To explain this detachment, XPS analyses were carried out, which are presented in the next section.

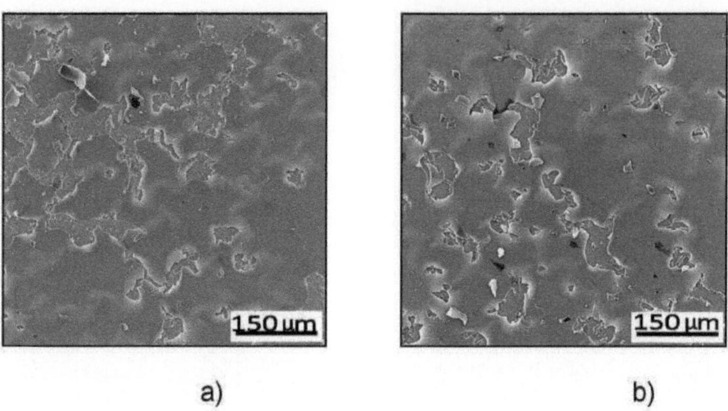

a)   b)

FIGURE 4 .48 SEM photomicrographs showing detachment of the functional AlN film deposited in Conditions 2 and 3 (Table 3.2) 200x: a) on apo M2 and b) on apo D6.

4.2.3.3. Characterization of Functional TiN and AlN Films by XPS

4.2.3.3.1. Functional Titanium Nitride Films

The Ti 2p spectrum for the samples with TiN films deposited on M2 and D6 steels (Condition 1) can be broken down into three components: the Ti 2p peak$_{3/2}$ with the lowest binding energy corresponds to TiN, the one with intermediate energy to Ti2O3 and the one with the highest energy to TiO2 (Figures 4.49 and 4.50).

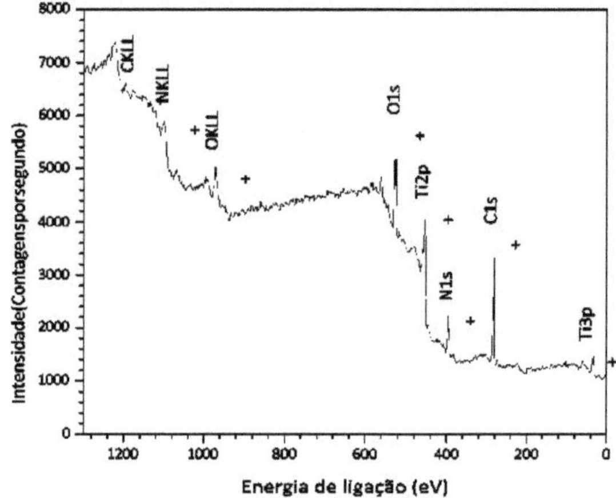

FIGURE 4 .49 Spectral curve obtained by XPS of the M2 steel sample with TiN film deposited in Condition 1 (Table 3.2).

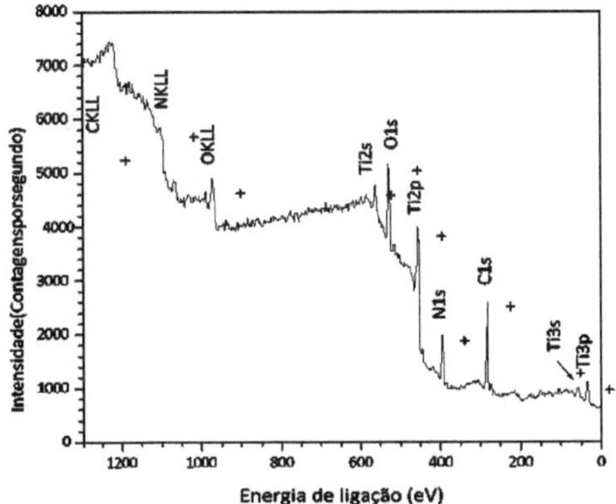

FIGURE 4 .50 Spectrum obtained by XPS of the apo D6 sample with TiN film deposited in Condition 1 (Table 3.2).

4.2.3.3.2. Functional Aluminum Nitride Films

All the AlN films deposited on M2 and D6 in Condition 2 peeled off completely. Therefore, it was not possible to analyze their surfaces by XPS.

In the samples with functional AlN films deposited on M2 and D6 (Condition 3 of Table 3.2), the presence of iron on the surface was detected, indicating that the aluminum nitride films did not completely cover the substrates in these two cases. This is the result of this film detaching from the surfaces of the substrates used. The Fe 2p spectrum for the interface between apo M2 (Figure 4.51) and the film corresponds to Fe2O3, while for the interface between apo D6 and the film (Figure 4.52), the Fe 2p3/2 peak was broken down into three components: the one with the lowest binding energy corresponds to metallic iron, the one with intermediate energy to FeO, and the one with the highest energy to Fe2O3. The presence of these last two compounds is due to the equipment used in this work (magnetron sputtering) not having a good vacuum and a cleaning system for the sputtered substrates (Item 3.2.3.2). The formation of these compounds on the surfaces of the substrates may have been the main cause of the detachment of the functional AlN films. Another factor is the high oxygen demand of aluminum, which may also have contributed to the detachment of these functional AlN films.

The Al 2p spectrum shows the presence of two components for the sample with AlN film deposited on D6 steel (Condition 3) and only one component for the samples with AlN film deposited on M2 steel (Condition 3) and with AlN film deposited on M2 and D6 (Condition 4) (Figures 4.51 to 4.54). AlN shows the Al 2p peak at 74.4 eV, which corresponds to the energy of Al2O3[108] . The peak for pure Al appears at 72.9 eV. The component with energy at 71.4 eV for the sample with the AlN film in D6 (Condition 3) may be associated with aluminum carbide (which may or may not be a chemically stoichiometric compound). Corroborating these statements, for the same sample, there is a carbon component at 283.7 eV, which corresponds to metallic carbide.

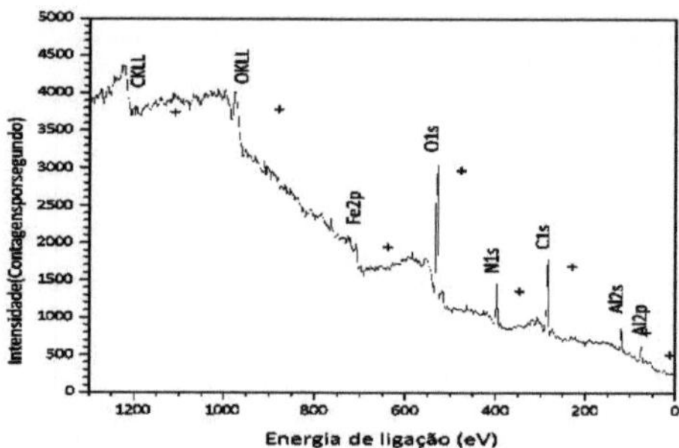

FIGURE 4 .51 Spectrum of steel sample M2 with AlN film
deposited in Condition 3 (Table 3.2), obtained by XPS.

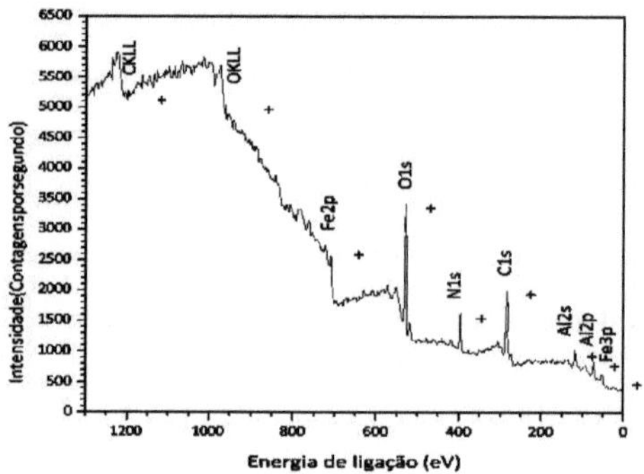

FIGURE 4 .52 Spectrum obtained by XPS of the apo D6 sample with AlN film deposited in Condition 3
(Table 3.2).

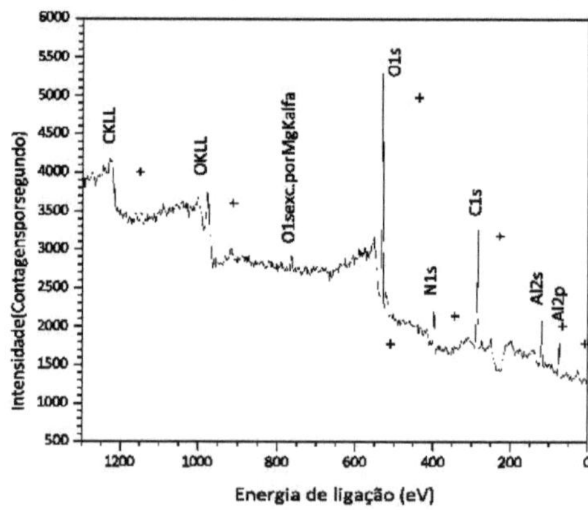

FIGURE 4 . 53 Spectrum of steel sample M2, obtained by XPS, with AlN film deposited in Condition 4
(Table 3.2).

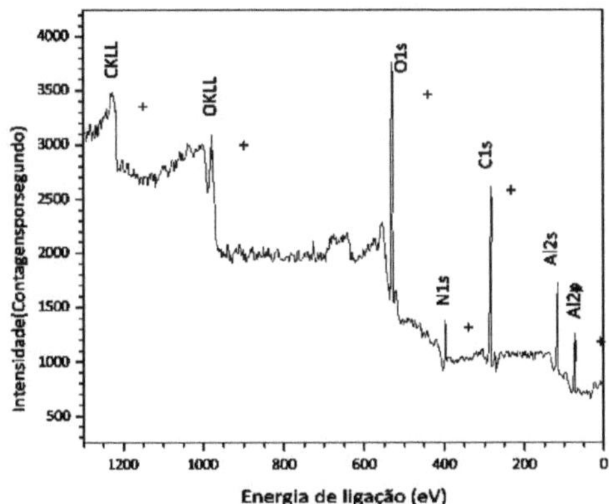

FIGURE 4 .54 XPS spectral curve of the apo D6 sample with AlN film deposited in Condition 4 (Table 3.2).

The binding energies of the main photoelectric peaks are shown in Table 4.8. The value in brackets represents the percentage of the component in the peak.

TABLE 4.8. Binding energies (in eV) from XPS spectra.

Sample	Bond energies (eV)					
	C 1s	The 1s	N 1s	Ti 2p3/2	Al 2p	Fe 2p3/2
TiN in M2 (Cond. 1)	282,6 (10%) 284,8 (68%) 286,6 (16%) 288,5 (6%)	529,8 (49%) 531,6 (36%) 533,3 (15%)	396,7 (70%) 399,1 (19%) 400,6 (11%)	455,3 (31%) 456,9 (27%) 458,4 (42%)	-	-
TiN in D6 (Cond. 1)	284,8 (89%) 286,9 (11%)	529,8 (47%) 531,5 (36%) 533,3 (17%)	396,7 (79%) 399,2 (21%)	455,3 (28%) 456,8 (25%) 458,4 (47%)	-	

AlN in M2 (Cond. 3)	284,8 (32%) 286,1 (51%) 288,1 (12%) 289,8 (5%)	529,6 (13%) 531,2 (32%) 533,0 (54%)	398,2 (47%) 399,9 (53%)	-	75,3	710,1
AlN in D6 (Cond. 3)	283,7 (4%) 284,8 (59%) 286,2 (20%) 287,9 (10%) 289,1 (7%)	529,2 (19%) 530,9 (38%) 532,6 (34%) 534,0 (9%)	397,9 (40%) 399,7 (60%)	-	71,4 (12%) 75,1 (88%)	706,1 (15%) 708,6 (33%) 710,2 (52%)
AlN in M2 (cond. 4)	284,8 (76%) 286,2 (12%) 288,6 (12%)	530,5 (23%) 531,9 (77%)	396,8 (80%) 399,4 (20%)	-	73,9	-
AlN in D6 (Cond. 4)	284,8 (66%) 286,4 (10%) 288,1 (11%) 289,9 (13%)	530,1 (22%) 531,7 (69%) 533,0 (8%)	396,7 (80%) 399,2 (20%)	-	74,0	-

The C 1s spectrum can be broken down into several components, ranging from two for the sample with TiN film on D6 (Condition 1) to five for the sample with AlN film on D6 (Condition 3). The component at exactly 284.8 eV, used as an energy reference, is associated with the C-C and C-H of hydrocarbons adsorbed on the surface. According to some authors[109] , C 1s at approximately 286 eV corresponds to C-OH, at approximately 288 eV, to C=O, and at approximately 289 eV, to carboxylic acid. The chemical component with the lowest binding energy corresponds to a metal carbide. In this case, the presence of these chemical components adsorbed on the surface is possible because it was not possible to clean the substrate surfaces adequately by sputtering, which would have greatly helped to eliminate these chemical components. The O 1s spectrum can be broken down into two (AlN film on M2 (Condition 4)), three (AlN film on M2 (Condition 3), AlN on D6 (Condition 4), TiN on M2 and D6 (Condition 1)) and four chemical components (AlN on D6 (Condition 3)). According to the same authors cited in the previous paragraph[109] , the O 1s peak with binding energy in the range 531.1 to 531.8 eV corresponds to C=O or carbonate, and in the range 532.3 to 533.3 eV, to C-OH. The metal oxides show the O 1s peak at lower binding energies, except for Al2O3, whose peak is at approximately 531 eV.

The N 1s spectrum shows two chemical components for almost all the samples, and three for the sample with the TiN film on M2 (Condition 1). The chemical component with the lowest binding energy corresponds to the metal nitride. The chemical component with the highest binding energy may be associated with nitrogen containing oxygen.

The composition of the samples, in atomic percentages, is shown in Table 4.9.

TABLE 4.9. Composition (atomic %) of XPS spectra.

Sample	Composition (atomic %)					
	C	O	N	Ti	Al	Fe
TiN in M2 (Cond. 1)	4,82	2,04	48,93	44,21	-	-
TiN in D6 (Cond. 1)	4,98	2,55	47,31	45,16	-	-
AlN in M2 (Cond. 3)	4,13	3,76	44,92		44,97	2,22
AlN in D6 (Cond. 3)	4,25	3,25	42,44	-	43,68	6,38
AlN in M2 (Cond. 4)	4,92	3,55	44,39	-	47,14	-
AlN in D6 (Cond. 4)	5,97	4,57	42,55	-	46,91	-

According to this analysis, the functional titanium nitride film deposited on apo M2 had a higher carbon concentration, lower oxygen concentration and a higher titanium and nitrogen concentration than the same film deposited on apo D6.

In the functional aluminum nitride films, the concentrations are very similar, with the exception of the concentration of oxygen, which is slightly higher, and nitrogen, which is lower, in the films grown on apo D6 (Condition 4) and the concentration of aluminum, which is higher in the films deposited on apo M2 (Condition 4). Aluminum nitride films have the highest oxygen concentration. This may be associated with the high avidity of this element with the oxygen present in the chamber due to the lack of a better vacuum, and on the surface of the substrates, due to the lack of cleaning by sputtering. The carbon concentrate found in all the films is due to possible contamination of these samples with the diffuser pump used in this deposition.

4.2.3.3.3 Partial conclusions on functional AlN films

TABLE 4.10. Characteristics of the deposited AlN functional films.

Type of film		Condido de deposido	Temperature (° C)	Adherence
	Type of substrate			

		2	114	terrible
AlN	M2	3	114	regular[0]
		4	135	regular[0]
	D6	2	114	terrible
		3	114	regular[0]
		4	135	regular[0]

(*)Good adhesion only for a few weeks.

The results obtained from the X-ray diffractograms show no influence of the type of substrate on the nucleation and growth of the functional aluminum nitride films. However, the higher deposition temperature of these films, associated with the absence of argon in the plasma (Condition 4), on the surface of D6 steel, shows the formation of a new preferential growth direction.

 SEM images show that the surfaces of the functional aluminum nitride films deposited in Condition 3 have fewer defects than the same films deposited in Condition 2. This is associated with the reduced presence of argon gas in the plasma under the conditions described in Table 3.2. This was proven in the deposition of this film in Condition 4 (in which the plasma was composed only of nitrogen) in which the film showed even fewer defects.

Subsequent SEM observations showed the total detachment of all the deposited aluminum nitride functional films.

XPS analysis shows the formation of AlN, but also of intermediate energy FeO and higher energy Fe_2O_3 and Al_2O_3 chemical compounds. The formation of these compounds on the surfaces of the substrates may have been the main cause of the detachment of the functional AlN films. The low vacuum combined with the high avidity of aluminum with oxygen present in the chamber, despite the presence of N2, and on the surface of the substrates contributed to these formations.

The total detachment of these films made it impossible to analyze them by AFM, RBS and 4-point bending tests of the film-substrate assembly.

The results of the XPS analysis showed that the carbon present in the functional AlN films did not form chemical compounds with the chemical components of the film and substrates used in this work. According to the Ellinghan diagrams for the formation of oxides, carbides and nitrides in these systems (Figures 4.55 and 4.56), the formation of aluminum oxide is the most likely to occur from a thermodynamic point of view. However, the experiments were carried out under conditions where the partial pressure of N2 was much higher than that of O2, thus favoring the re-formation of aluminum nitride[137,138] .

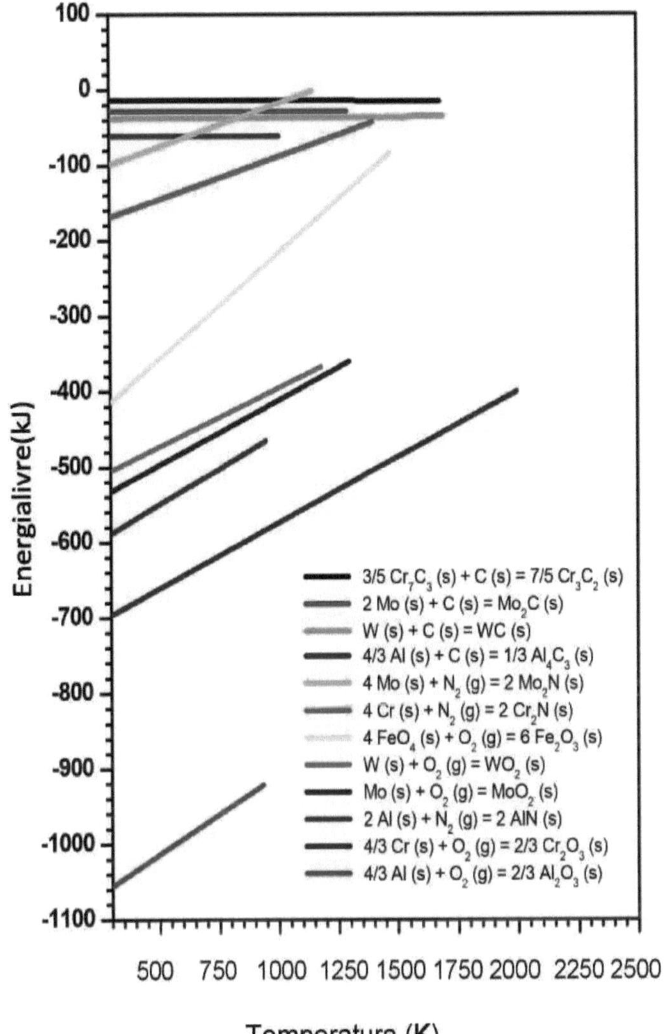

FIGURE 4 .55 Ellingham diagram for the reappearance of oxide, nitride and carbide formations on AlN films deposited on the surfaces of M2 fast apo substrates[137] .

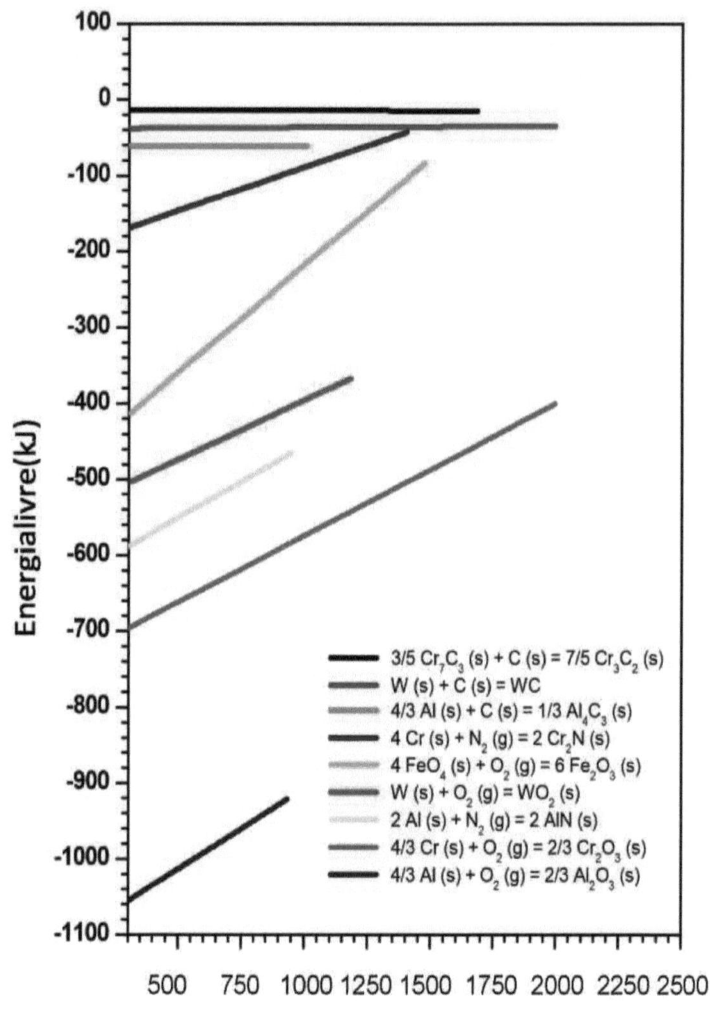

FIGURE 4 .56 Reappearance of oxide, nitride and carbide formations, shown by Ellingham diagram, on AlN films deposited on the surfaces of apo tool substrates[D6] [137]..

4.2.3.4. Characterization of Functional TiN Films by AFM

The surface roughness of the titanium nitride films deposited on the surfaces of M2 and D6 steels in Condition 1 (Table 3.2) was mapped using the atomic force microscopy (AFM) technique. The observations resulting from these mappings are shown in Figure 4.57.

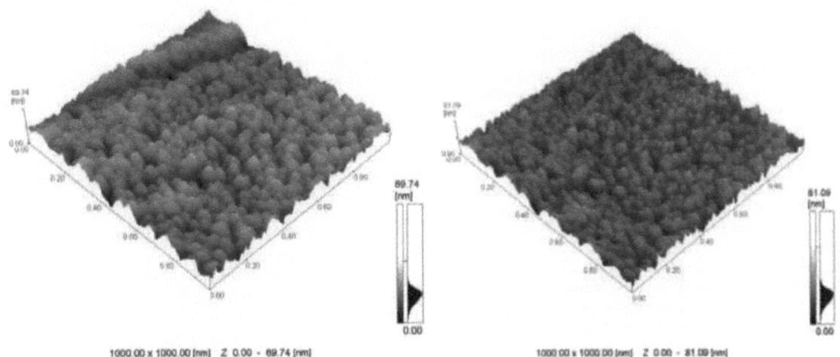

FIGURE 4.57. AFM images of the surfaces of the TiN functional film deposited under Condition 1 (Table 3.2): a) on M2 steel and b) on D6 steel.

Observations of the surface morphology obtained by AFM indicate that the TiN functional film deposited on the surface of M2 steel is slightly rougher on the surface than the same film deposited on D6 steel, with roughness values of 9.223 nm and 9.151 nm, respectively. Both surfaces have a homogeneous microstructure with nanocrystals of similar size and shape (Figures 4.57 a and b). The low temperature
used in this deposition seems to have contributed to the good nanostructured formation of this film. As already discussed in Items 4.2.1.3 and 4.2.1.4, TiN films deposited at a lower temperature present a better deposition condition.

4.2.3.5. Characterization of Functional TiN Films by RBS

To determine the atomic composition and atomic depth concentration profile of the titanium nitride functional films deposited under the conditions specified in Table 3.2, analyses were carried out using the Rutherford backscattering spectroscopy (RBS) technique. The spectra resulting from the RBS analyses for the samples of M2 high-speed steel and D6 tool steel with titanium nitride films deposited are shown in Figures 4.58 and 4.59. In these spectra, the red curve represents the experimental values, while the black curve represents the simulation for TiN on M2 and D6 steels using the RUMP program[134].

The spectrum obtained by RBS for the titanium nitride functional film deposited on M2 steel is shown in Figure 4.58. The concentration values for each layer obtained in the RUMP simulation (experimental curve) are listed in Table 4.11.

The concentration values for each layer were found to be close to the experimental curve found by RBS. The thickness values for each layer were 50 nm for each functional layer and 80 nm for the intermediate titanium layer. The intermediate layer has a higher titanium concentration, as expected. However, it also has concentrations of argon, oxygen and carbon. In each functional layer deposited, according to the RUMP analysis (Table 4.11), there was an increase in titanium, a decrease in nitrogen and a concentration of carbon. The presence of carbon, depending on the deposition time, may be associated with contamination from the diffusion pump of the deposition equipment. On the other hand, there was a presence of oxygen in the layers, which decreased over the course of the deposition. The concentrations of argon in all the layers are very similar. Argon does not react with any chemical element and must therefore be in solid solution. The presence of argon inclusions does not seem to influence the mechanical strength of this film.

By simulating the TiN functional film/Ti film/M2 substrate system, the following elements are present: 34.73% Ti, 11.58% Ar, 30.90% N, 0.21% O, 18.23% Fe and 4.35% C. The values obtained in the simulation are approximately 200 nm for the thickness of the TiN functional film on M2 and the titanium and nitrogen concentrations shown in Table 4.11.

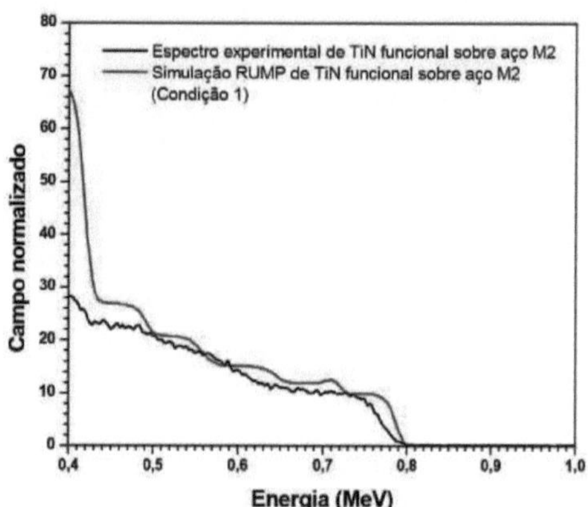

FIGURE 4 .58 Spectra obtained by experimental and simulated RBS for the TiN functional film deposited in Condition 1 on apo M2 (Table 3.2).

TABLE 4.11. Atomic concentrations of the chemical elements in each layer of the TiN functional film deposited on M2 steel (Condition - Table 3.2).

Ti/N layer (%)	Ti (%)	Air (%)	N (%)	O (%)	C (%)	Thickness (nm)
5ocamada 50 / 50	54,67	17,39	21,05	2,43	4,01	50
4 layer 40 / 60	49,50	19,56	23,38	2,69	4,87	50
3-layer 30 / 70	41,09	21,73	26,31	4,31	6,56	50
2nd layer 20 / 80	35,58	23,91	28,94	4,39	7,18	50
leased 100 / 0	70,14	17,39	0,46	4,39	7,62	80
M2 high speed steel substrate						

The analysis of the spectrum obtained by RBS for the titanium nitride functional film deposited on D6 steel

(Condition 1) is shown in Figure 4.59. The concentration values of the chemical elements present in each layer obtained in the simulation (experimental curve) are shown in Table 4.12.

This analysis shows thickness values of 50 nm for each functional layer and 80 nm for the intermediate titanium layer. According to these analyses, the titanium nitride functional film deposited on D6 steel also shows oxygen and carbon contamination. As expected, the intermediate layer has a higher concentration of titanium. However, concentrations of argon, oxygen and carbon are also observed. This analysis proves that, with each functional layer deposited, there is a small increase in titanium and a decrease in nitrogen and argon. As argon is inert and therefore doesn't react with any chemical element, it may be in solid solution, but it doesn't seem to have affected the mechanical strength of the film. On the other hand, oxygen is also present in the layers.

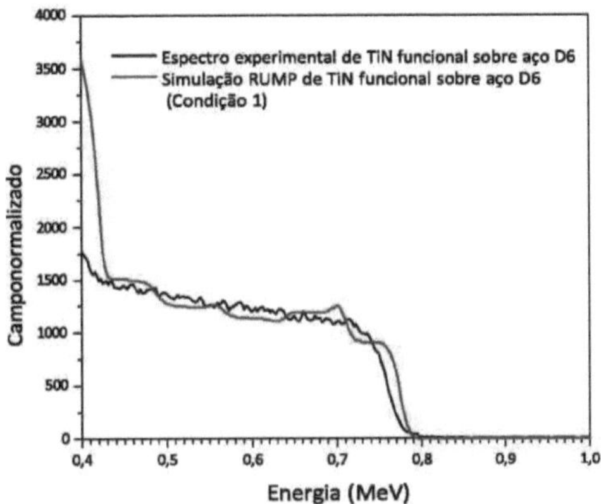

FIGURE 4 .59 Experimental and simulated RBS spectra for the film
functional TiN deposited on apo D6 (Table 3.2).

TABLE 4.12. Concentrations of the chemical elements in each layer of the TiN functional film deposited on apo D6 (Condition 1 - Table 3.2).

Ti / N layer (%)	Ti (%)	Air (%)	N (%)	O (%)	C (%)	Thickness (nm)
5° layer 50 / 50	53,88	17,39	21,05	3,43	4,25	50
4° layer 40 / 60	48,64	19,57	23,39	3,68	4,72	50
3° layer 30 / 70	42,41	21,39	26,31	3,74	6,15	50
2° layer 20 / 80	36,17	23,91	28,95	4,39	6,58	50

1° layer 100 / 0	69,03	17,39	0,87	4,39	8,32	80
D6 tool support substrate						

The simulation of the TiN functional film/Ti film/D6 substrate system shows that 35.79% Ti, 10.68% Ar, 30.20% N, 0.18% O, 16.67% Fe and 6.48% C are present. The values obtained in the simulation were approximately 200 nm for the thickness of the TiN functional film deposited on D6 and the titanium and nitrogen concentrations shown in Table 4.12.

4.2.3.6. RUMP Simulation for Functional TiN Films

Simulations were carried out with the most suitable energy and 0 parameters, obtained in Item 4.2.I.5.2., to analyze only titanium and nitrogen in the TiN functional film on M2 and D6. Considering a model starting from the surface to the substrate with a titanium and nitrogen concentration in the sequence (according to beam detection) of 50%/50%, 40%/60%, 30%/70%, 20%/80%, respectively, 100% Ti as the intermediate layer and substrate composition M2 and D6 (as described in Item 4.1.2), simulation spectra of these functional films were obtained using the ideal values, found in accordance with Item 4.2.1.5.2, for an energy of 1.1 MeV and 0 = -60° (Figures 4.60 and 4.61).

Figure 4.60 shows a simulation spectrum of a TiN functional film deposited on M2 support. It can be seen that with the use of the new energy and 0 parameters, it is possible to analyze titanium throughout the depth of the TiN functional film, showing the difference in composition and without the interference of the presence of heavier elements. However, it is still difficult to analyze nitrogen.

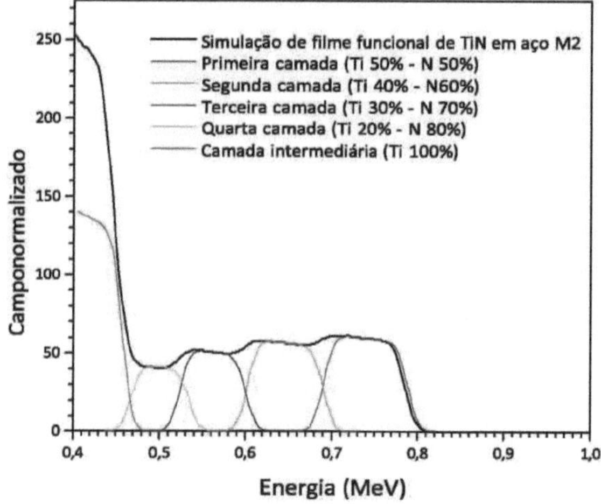

FIGURE 4.60 Simulation of a functional TiN film on M2 substrate considering an energy of 1.1 MeV, 0 = -60° , titanium and nitrogen composition in the sequence of 50/50, 40/60, 30/70 and 20/80, respectively, an intermediate layer of 100% Ti and M2 substrate with composition as described in Item 4.1.2.

The spectrum obtained by simulating the TiN functional film deposited on apo D6 shows a simulated spectrum (Figure 4.61) very similar to that of apo M2 (Figure 4.60). It is also possible to analyze the titanium element throughout the depth of the TiN functional film, showing the difference in composition and without the interference of heavier elements.

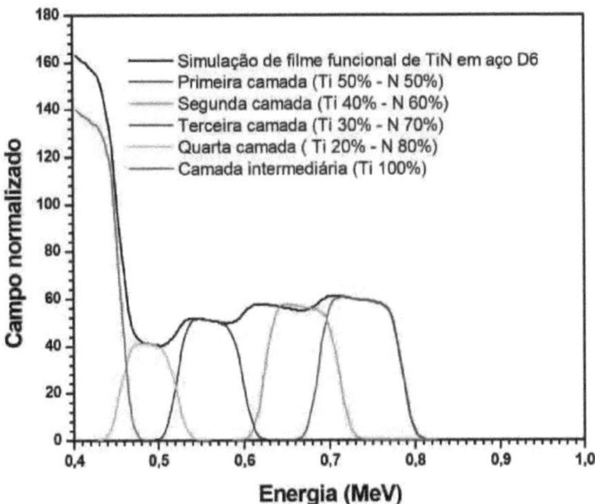

FIGURE 4.61 RUMP simulation curves of a functional TiN film on D6 steel considering an energy of 1.1 MeV, 0 = -60° , titanium and nitrogen composition in the sequence of 50/50,   40/60, 30/70 and 20/80, respectively, and an intermediate layer of 100% Ti and substrate D6 with the composition described in Item 4.1.2.

4.2.3.7. Bending Tests on Functional TiN Films

The M2 and D6 steel substrates with polished surfaces and functional titanium nitride films deposited in Condition 1 (Table 3.2) were subjected to the 4-point bending test. Photomicrographs of the samples prepared according to the procedures adopted in Item 3.2.4.10 are shown (Figures 4.62 and 4.63). Despite an adequate search, no standard or reference was found in the literature that established the parameters for making specimens and/or conditions for bending tests for substrates with deposited films. It was therefore necessary to adapt the ASTM E855 - 90 standard for the tests on the samples prepared in this work, as described in Item 3.2.4.10.

In both cases, during the 4-point bending test, shortly after the maximum load was applied, the titanium nitride functional film showed the formation of "scales" indicating the localized loss of its adhesion to the substrates of the M2 and D6 steels. This "flaking" was more pronounced in the region close to the load application points in the 4-point bending tests (Figures 4.62a and b and 4.63a and b). Very small cracks occur in the center of the two steel samples, with a minimization of the "flaking" effect (Figures 4.62c and d and 4.63c and d). The two types of steel with titanium nitride functional films show very similar "flaking" with good distribution.

Stress versus strain curves were obtained for both tests.

The stress versus strain curves for the 4-point bending test (Figures 4.64 and 4.65) show consistent behavior. For deformation values of around 13.80 %, in the case of this film deposited on M2 steel, and on D6, around 15.10 %, the curve tends to show the behavior of the substrate material. This behavior can be related to the rupture and/or detachment of the functional titanium nitride film. The stresses associated with these changes in the behavior of the curves, both for the film deposited on apo M2 and apo D6, are approximately 1.36 and 1.31 GPa, respectively.

These results indicate that, although these films were deposited on different substrates, the results are very similar and the film breaks away from the surface of the substrates after close to the maximum tension.

In future work we intend to carefully study the behavior of these curves as a function of test speed and film thickness. It will also be necessary to study ductile and brittle films. The ultimate goal should be to establish a standard for tests to determine the adhesion of films to substrates.

Unfortunately, it was not possible to characterize the functional AlN films obtained in this work, as they completely detached from the surfaces of the respective substrates.

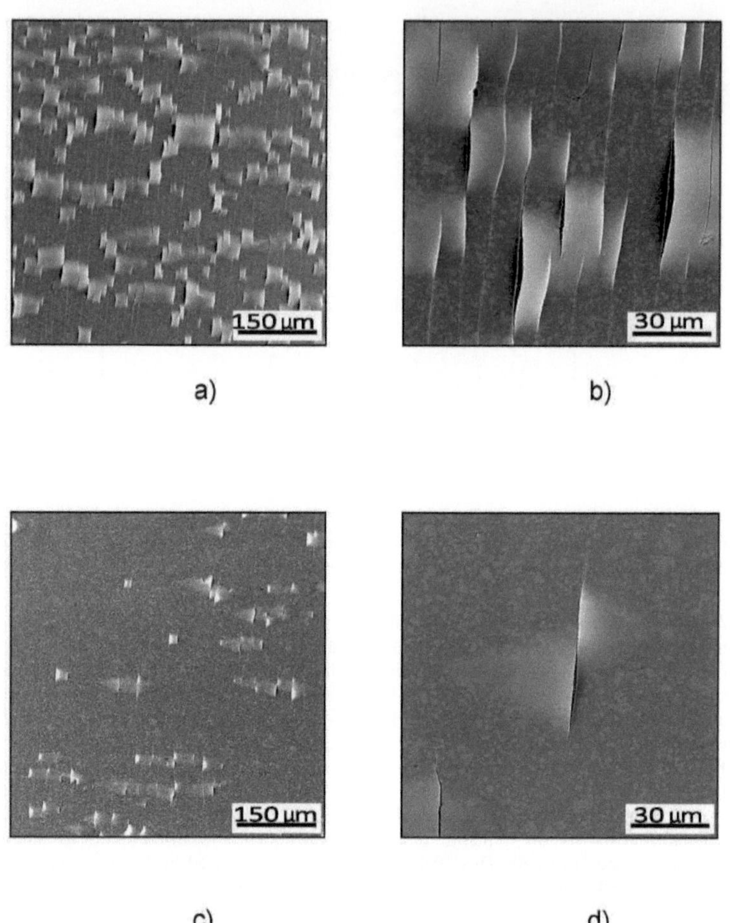

FIGURE 4 .62 SEM photomicrographs of the polished surface of M2 steel with TiN film deposited in Condition 1 (Table 3.2): a) and b) region of the load application on the TiN functional surface; c) and d) center of the sample on the TiN functional surface.

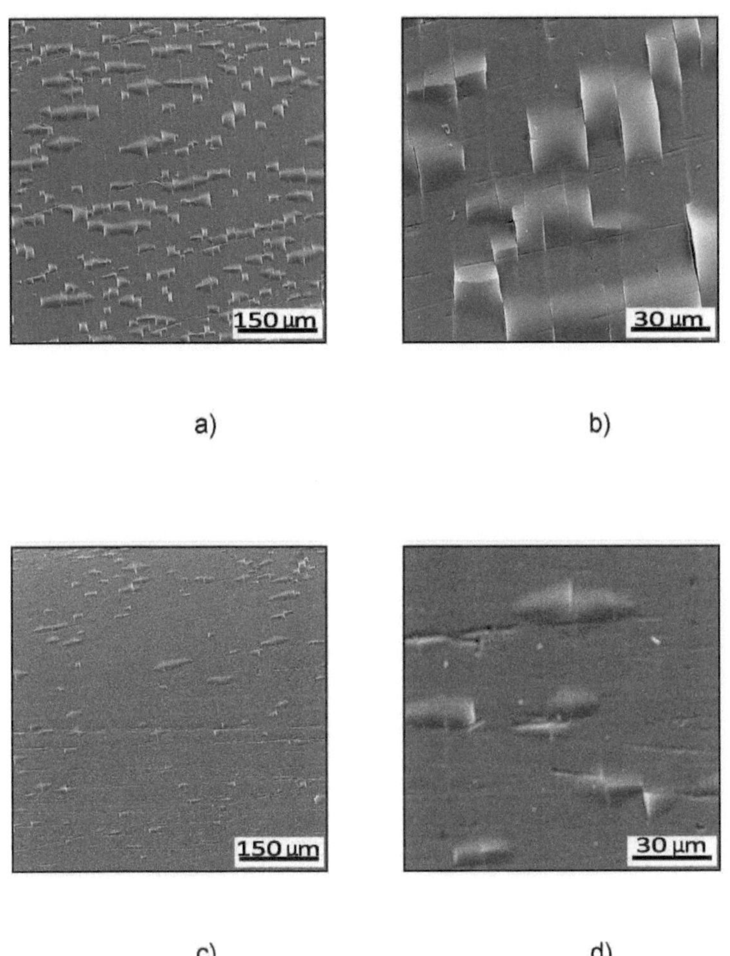

FIGURE 4 .63 Photomicrographs of the surface of D6 steel with TiN film deposited in Condition 1 (Table 3.2), obtained in SEM: a) and b) region of the load application on the TiN functional surface; c) and d) center of the sample on the TiN functional surface.

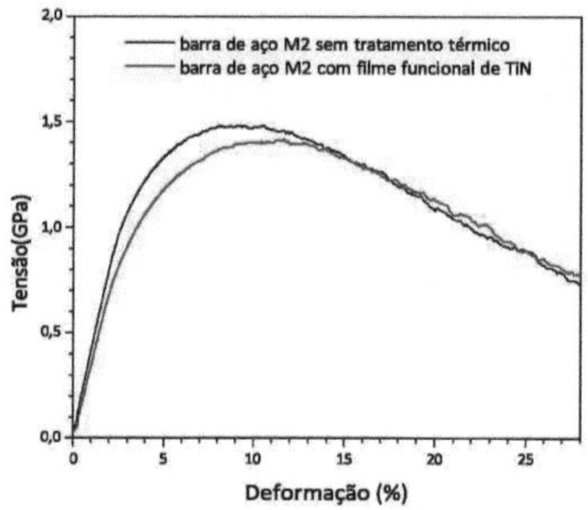

FIGURE 4.64 Stress versus strain curves obtained from
4-point bending tests for apo M2 samples with
TiN functional films deposited in Condition 1.

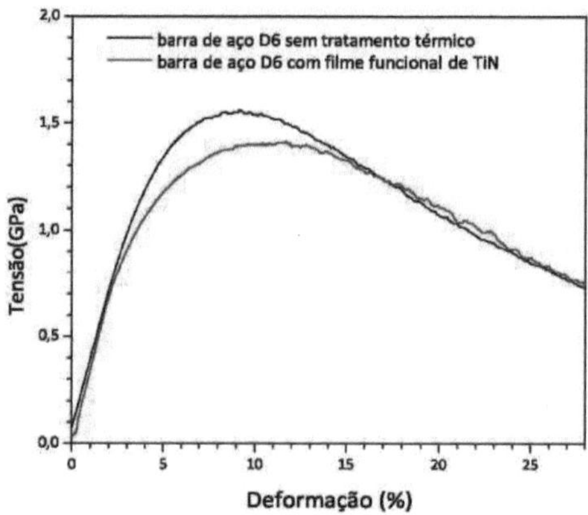

FIGURE 4 .65 Curves obtained from 4-point bending tests for D6 steel samples with TiN functional films
deposited in Condition 1 (Table 3.2).

4.2.3.8. Partial Conclusions on Functional TiN Films

The X-ray diffractograms showed the formation of TiN on all the substrates deposited in Condition 1. The metastable phase Ti2N(202) was formed on the apo M2 substrate.

The surfaces of these films deposited on M2 and D6, observed by SEM, showed shallow scratches and homogeneous pore size distribution.

According to AFM observations for this functional TiN film, and in line with results already shown for TiN films (Item 4.2.1.3), the low temperature used in this deposition seems to have contributed to the good nanostructured formation of this film.

XPS analysis showed that in these functional films there was the formation of chemical components with the lowest binding energy corresponding to TiN, intermediate energy to Ti2O3 (metastable compound) and the highest energy corresponding to TiO2. The presence of carbon and oxygen was also found in this film, but this does not seem to have influenced the quality and adhesion of these films, since they must be present in the form of solid solutions in the films.

According to the Ellinghan diagrams for the formation of oxides, carbides and nitrides in these systems (Figures 4.66 and 4.67), the temperatures used to deposit the TiN films were not sufficient for the formation of titanium oxide, with the re-formation of chromic oxide being the most likely to occur from a thermodynamic point of view[138] . However, the films were deposited under conditions in which the partial pressure of N2 was much higher than that of O2, thus favoring the titanium nitride formation reaction .[137]

Observations by RBS showed a thickness of 200 nm for the functional layers, which is very close to the 250 nm estimated by the equipment. As already shown by XPS, the RBS technique showed a concentration of carbon and oxygen and a small amount of argon in these deposited functional films. In the last functional layer deposited, the least oxygen contamination seems to have occurred.

RUMP simulations, using the new parameters obtained in Item 4.2.1.5.2, showed that it was possible to observe the titanium element throughout the depth of the TiN functional film without interference from the presence of heavier elements, showing the difference in composition. On the other hand, it is still difficult to analyze the nitrogen element by RBS.

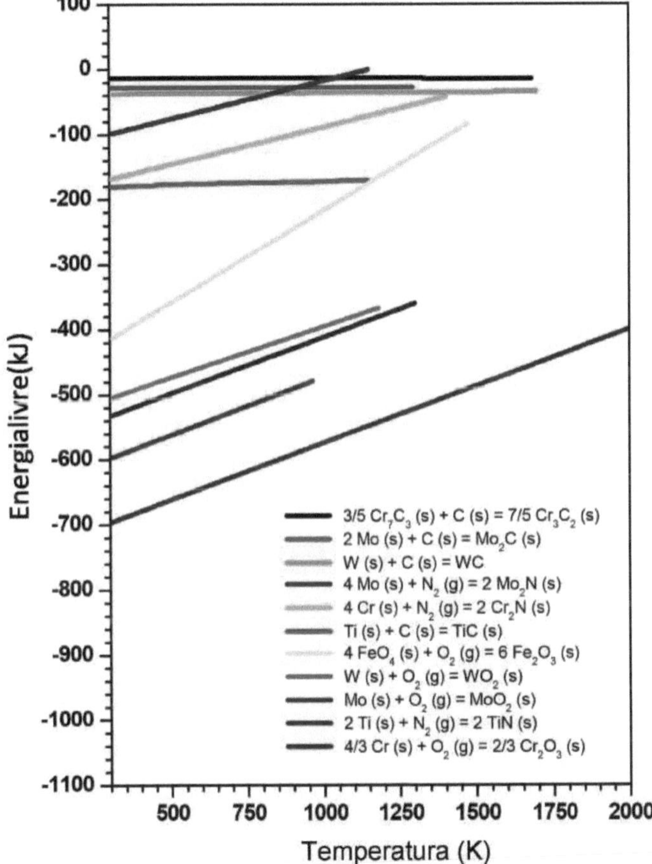

FIGURE 4 .66 Ellingham diagram showing the reappearance of oxide, nitride and carbide formations on TiN films deposited on the surfaces of M2 fast apo substrates[137,138] .

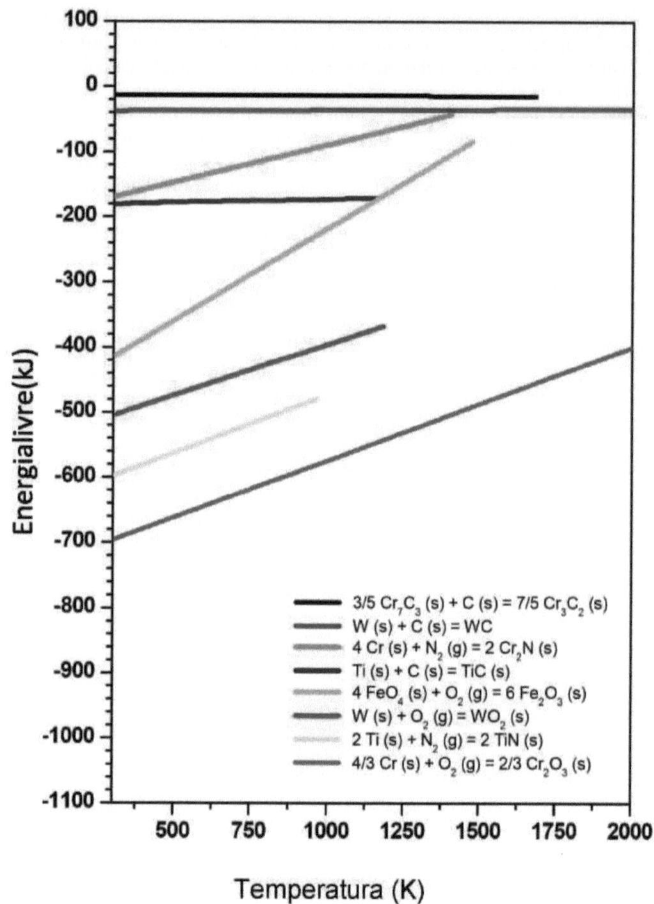

FIGURE 4 .67 Ellingham diagram showing the reappearance of oxide, nitride and carbide formations on TiN films deposited on the surfaces of apo tool substratesD6 [137,138]..

The results of the 4-point bending tests indicated excellent adhesion of these films, showing values of tensile strength and/or detachment of these films deposited after M2 and D6 of approximately 4.12 and 4.06 GPa, respectively.

4.3 General discussion and conclusions

According to the results shown in Table 4.13, with the lower deposition temperature it was possible to obtain TiN films deposited after M2 and D6 with better film quality results, while the AlN film deposited at a lower temperature, despite the good film quality, did not show adequate adhesion. Thus, unlike the AlN films, the TiN films obtained better deposition results at lower temperatures.

TABLE 4.13. Comparison between the TiN and AlN films deposited.

Type of film	Type of substrate	Deposition temperature (° C)	Movie quality	Quality of Adherence

	M2	220	good	good
TiN	D6	220	good	good
	M2	450	regular	good
	D6	450	terrible	good
AlN	M2	143	good	terrible(*)
	D6	143	good	terrible(*)

(*) Immediately after deposition, the films detached from the surface of the substrates.

The function of the intermediate titanium film used to deposit the TiN films was of great importance in terms of adhesion. Even though the TiN films grown at a temperature of 450° C on apo D6 surfaces showed detachment near the surface of the film, they also showed good adhesion at the interface. On the other hand, the intermediate aluminum film used in the AlN depositions failed to create conditions at the interface for significant adhesion. The biggest problem with these AlN films may be related to the contamination of the aluminum by oxygen present in the deposition chamber, combined with its great avidity for this gas.
All the TiN films deposited showed significant adhesion.

TABLE 4.14. Comparison of the quality of functional TiN and AlN films.

Film type	Type of substrate	Deposit condition	Movie quality	Quality of Adherence
TiN	M2	1	great	great
	D6		great	great

AlN	M2	2	good	terrible
		3	good	regular^0 *)
		4	good	regular^0 *)
	D6	2	good	terrible
		3	good	regular^0 *)
		4	good	regular^0 *)

(*)Adherence was initially good, but after a few weeks it began to show signs of detachment.

The titanium nitride and aluminium nitride functional films apparently formed quite satisfactorily on the surface of M2 and D6. However, the carbon and oxygen contaminants had a greater influence on the adhesion quality of the AlN functional films than the TiN functional films. For proper deposition of this functional aluminum nitride film, equipment is needed that contains a pump that works as cleanly as a molecular turbo and that can reach higher vacuum pressures. It also needs to contain a system for cleaning the substrates by argon sputtering. In this way, it is hoped that it will be possible to produce functional AlN films without the presence of these contaminants and with good adhesion.

As can be seen in Table 4.14, the functional AlN films peeled off after a few weeks, unlike the case of the AlN films (Table 4.13) where the peeling was almost instantaneous as soon as the samples were removed from the deposition chamber. This fact shows the main advantage and importance of this new film deposition technique, called in this work functional films or engineered interface, which can actually dampen the intrinsic stresses in the region close to the interface in order to increase film adhesion. The only disadvantage of this new form of film deposition is the need for a chamber that contains minimal equipment such as: cleaning the substrates by sputtering, a pump that works cleanly (molecular turbo) and that can achieve larger voids, reducing the possibility of contaminants such as carbon and oxygen. Another need from a technological point of view is to have equipment that can deposit these thicker films with good layer formation quality. The magnetron sputtering technique does not seem to meet this requirement.

The functional TiN films also showed significant improvements over the TiN films, which could be seen mainly by comparing the behavior of these films after the 4-point bending test. After the bending test, the functional TiN films showed the formation of "scales" on their surfaces, while the TiN films only formed cracks. The formation of these "scales" is related to the greater adhesion of the film to the substrate, where the interface region has a gradual variation[139].

Despite realizing the importance of nano-hardness measurements for this work, it was not possible to carry them out because the equipment was undergoing maintenance.

CONCLUSIONS

5.1. General conclusions

The aim of this work was to develop a new technique for producing functional films with a designed interface, whose chemical composition varies gradually according to the thickness of the film.

TiN films and TiN functional films remained adherent to the surfaces of M2 and D6. The functional TiN films showed the best adhesion results, proving the importance of relieving mechanical stresses near the film-substrate interface region.

The intermediate titanium film influenced the increase in adhesion of the TiN films deposited on the substrates after M2 and D6. However, the aluminum film did not show this behavior, probably due to the oxidation of the substrate surface by the oxygen present in the vacuum chamber during the deposition stage. Although the functional AlN films detached from the substrates, they remained longer on the surfaces after M2 and D6 when compared to the AlN films deposited directly on intermediate Al films.

The adhesion of the functional AlN films was low and that of the AlN films even lower, due to: i) contamination of the deposition chamber with oxygen and ii) the lack of a system for cleaning the surfaces of the steel substrates by argon sputtering, coupled to the deposition chamber.

The studies carried out in this research work have resulted in the optimization of the chemical analysis techniques by RBS and simulation by RUMP and in obtaining greater precision in the results of the analysis of these types of films. The use of these analyses in the chemical characterization of functional films is original and will be of great importance in the investigation of functional film interfaces.

5.2. Proposals for complementary research work

The results and conclusions shown in this thesis lead to the suggestion of several issues that should be investigated in future work:

- Study of the nano-hardness behavior of these functional films and
- Systematic investigation of the behavior of 4-point bending curves as a function of test speed and film thickness.

BIBLIOGRAPHICAL REFERENCES

Conrad, J.R.; Dodd, R.A.; Worzala, F. J. Plasma source ion implantation: a new manufacturing process for surface
modification of materials. In:NSF Grantees Conference on Production Research and Technology, Society of Manufacturing Engineers, 15., 1989. Proceedings... [S.l.]: [S.n.], 1989. p.1-9.

Nono, M.C.A.; Ueda, M.; Stellati, C.; Barroso, J.J. Ion implantation in materials by plasma immersion. In: Congresso Brasileiro de Engenharia e Ciencia dos Materiais, 11., 1994, Aguas de Sao Pedro, SP. Proceedings... [S.l.]: [S.n.], 1994.

Rej, D.J. Plasma immersion ion implantation (PIII). In: Glocker, D.A.; Shah, S.I. (ed). Handbook of Thin Film Process Technology. Bristol: Institute of Physics Publishing, 1996.

Rie, K.T.; Menthe, E.; Mattews, A.; Legg, K.; Chin, J. Plasma surface engineering of metals. MRS Bulletin, v.21, p.46-51, 1996

Sioshansi, P. Surface modification of industrial components by ion implantation. Material Science Engineering, v.90, p.373-383, 1987

Nono, M.C.A.; Stellati, C.; Ueda, M.; Barroso, J.J.; Corat, E.J., Conrad, J.R.; Shamin, M.M.; Fetherston, P.; Kumar, K. Modification of metal surfaces by ion implantation and combined techniques involving ion deposition and implantation to increase the adhesion of CVD diamond films. In: Congresso Brasileiro de Aplicações de Vacuo na Industria e na Ciência, 16., 1997, Petropolis, RJ. Proceedings... Petropolis: [S.n.], 1997.

Perepezko, J.; Brewer, L.; Schaefer, R. Principles underlying coatings and surface modification science. Material Science Engineering, v.70, p.9-22, 1985.

Mattox, D.M.; Greene, J. E.; Buckley, D. H.; Somorjai, G. A. Properties of coated and modified surfaces. Material Science Engineering, v.70, p.79-89, 1985.

Hirvonen, J.K. Ion beam assisted thin film deposition. Material Science Report, v.6, p.215-276, 1991.

Nono, M.C.A.; Corat, E.J.; Ueda, M.; Stellati, C.; Barroso, J.J.; Conrad, J.R.; Shamin, M.M.; Fetherston, P.; Sridharan, K. Surface modification on 304 SS by plasma immersed ion implantation to improve the adherence of a CVD diamond film. Surface Coating Technology, v.112, p. 95-298, 1999.

Nono, M.C.A., Conrad, R.; Shaminm, M.; Fetherston, P.; Sridharan, K. Surface modification of materials by ion implantation and combined techniques involving film deposition and ion implantation to enhance film adhesion on metallic substrates. In: Brazilian Congress of Materials Science and Engineering, 13, 1998. Proceedings... [S.l.]: [S.n.], 1998.

Nono, M.C.A.; Corat, E.J.; Conrad, R.; Shaminm, M.; Fetherston, P.; Sridharan, K. Nucleation and growth of CVD diamond on 304 SS surface modified by plasma immersed ion implantation. In: International Conference on Advances in Surface Engineering, 5.; International Seminar on Surface Engineering, 2., 1998, Hotel Deville, Guarulhos, SP. Proceedings... Sâo Paulo: [S.n.], 1998.

Winters, H.F.; Chang,   R.P.H.; Mobab, C.J.; Evans, J.; Thornton, J.A.; Yasuda, H.
Coatings and surface modification using low pressure nonequilibrium plasmas. Material Science Engineering, v.70, p.53-77, 1985.

Picraux, S.T. Ion implantation metallurgy. Physics Today, v.37, p.38-44, Nov. 1984.

Hochman, R.F. Ion implantation. American Society for Metals, v.5, p.422-426, 1980.

Mantese, J.V.; Brown, I.G.; Cheung, N.W.; Collins, G.A. Plasma-immersion ion implantation. MRS Bulletin, v.21, p. 5256, 1996.

Collins, G.; Rej, D. J. Plasma processing of advanced materials. MRS Bulletin, v.21, p.26-31, 1996.

Conrad, J.R.; Radtke, J.L.; Dood, R. A.; Worzala, F. J.; Tran, N. C. Plasma source ion implantation technique for surface modification of materials. Journal of Applied Physics, v.62, n.11, p.591-96, 1987.

Zou, X.P.; Kang, E.T.; Neoh, K.G.; Cui, C.Q.; Lim, T.B. Surface modification of poly(tetrafluoroethylene) films by plasma polymerization of glycidyl methacrylate for adhesion enhancement with evaporated copper. Polymer, v.42, p.64096418, 2001.

Zhang, M.C.; Kang, E.T.; Neoh, K.G.; Tan, K.L. Surface modification of aluminum foil and

PTFE film by graft polymerization for adhesion enhancement. Colloids and Surfaces, v.176, p.139-150, 2001.

Vieira, R.A. Study of surface modifications of 304 stainless steel with polymeric and titanium films to improve the adhesion of CVD diamond films. 2000. Dissertation (Master's Degree in Materials Engineering) - Lorena School of Chemical Engineering, Lorena, SP, 2000.

Avalos, H.S.N. Deposition of metallic and TiN thin films by electric arc discharge in vacuum. 1995. Dissertation (Master's Degree in Engineering) - Guaratinguetá School of Engineering,São Paulo State University ,Guaratingüetá, SP, 1995.

Nono, M.C.A.; Gomes, G.F.; Vieira, R.A.; Teixeira, J.C.; Algatti, M.A. Modifications of material surfaces
metal by ion implantation and film deposition under development at INPE. In: Congresso Brasileiro de
Materials Science and Engineering, 14., 2001, Águas de Sâo Pedro, SP. Proceedings... 2001. 1 CD-ROM. p.45301-45316.

Bell, T.; Mao, K.; Sun, Y. Surface engineering design: modelling surface engineering systems for improved tribological performance. Surface Coating Technology, v.108/109, p. 360-368, 1998.

Davis, J.R. Surface engineering of specialty steels. Surface Engineering, v.5, p.762-775, 1994.

Celis, J.P.; Drees, D.; Huq, M.Z.; Wu, P.Q.; De Bonte, M. Hydrid process - a versatile technique and coating needs. Surface Coating Technology, v.113, p. 165-181, 1999.

Brown, I.G.; Anders, A.; Dickinson, M.R.; Macgill, R.A.; Monteiro, O.R. Recent advances in surface processing with metal plasma and ion beams. Surface Coating Technology, v.112, p. 271-277, 1999.

Suchentrunk, R.; Fuesser, H.J.; Staudigl, G.; Jonke, D.; Meyer, M. Plasma surface engineering - innovative processes and coating systems for high-quality products. Surface Coating Technology, v.112, p.351-357, 1999.

Lewis, B.; Anderson, J.C. Nucleation and growth of thin films. New York: Academic Press, 1978.

Sikola, T.; Spousta, J.; Ceska, R.; Zlamal, J.; Dittrichova, L.; Nebojsa, A.; Navratil, K.; Rafaja, D.'Zemek, J.; Perina, V. Deposition of metal nitrides by IBAD. Surface Coating Technology, v.108, p. 284-291, 1998.

Vieira, R.A.; Nono, M.C.A.; Mattos, M.L.B. SEM and EDS investigations of the Ti film-steel substrate interface. Acta Microscopica, Supplement A, v.B, p.309310, 1999.

Vieira, R.A.; Nono, M.C.A. Interface studies of CVD diamond- titanium and titanium-304 ss obtained by hybrid coating process. In: Congresso Brasileiro de Engenharia e Ciencia dos Materiais, 14., 2001, Águas de Sao Pedro, SP. Proceedings...
2001. 1 CD-ROM. p.44601-44613.

Nono, M.C.A.; Vieira, R.A.; Gomes, G.F. CVD diamond nucleation on 304 ss surfaces modified by carbon and nitrogen ion implantation. Acta Microscopica, Supplement A, v.9, p.277-278, 2001.

Vieira, R.A.; Nono, M.C.A. Nucleation and growth of CVD diamond films on titanium intermediate layers deposited on 304 SS. Acta Microscopica, Supplement A, v.9, p.279-280, 2001.

Vieira, R.A.; Nono, M.C.A. Microsctructure and interface analyses of TiN thin films deposited on M2 speed steel substrate by cathodic arc technique. Journal of Metastable and Nanocrystalline Materials, Switzerland, v. 20-21, p. 654-658, 2004.

Vieira, R.A.; Nono, M.C.A. Surface analyses of TiN thin films grown on 304 stainless steel by cathodic arc technique. Acta Microscópica, v. 12, p. 173-174, 2003.

Bouzakis, K.D.; Vidakis, N.; Leyendecker, T.; Erkens, G.; Wenke, R. Determinationof   the fatigue properties of
multilayer PVD coatings on various substrates, based on the impact test and its FEM simulation. Thin Solid Films, v.308/309, p.315-322, 1997.

Rother, B.; Jehn, H.A.; Gabriel, H.M. Multilayer hard coatings by coordinated substrate rotation modes in industrial PVD deposition systems. Surface Coating Technology, v.86/87, p.207-211, 1996.

Bromark, M.; Larsson, M.; Hedenqvist, P.; Hogmark, S. Wear of PVD Ti/TiN multilayer coatings. Surface Coating Technology, v.90, p.217-223, 1997.

Carvalho, N.J.M.; Dehosson, J.T.M. Microstructure investigation of magnetron sputtered WC/C coatings deposited on steel substrates. Thin Solid Films, v.388, p.150159, 2001.

Hubler, R.   Ionbeam   mixing of Ti-TiN multilayers for tribologicalandcorrosion   protection. NIM B - Beam interactions with Materials & Atoms, v.175/177, p.630-636, 2001.

Duck, A.; Gamer, N.; Gesatzke, W.; Griepentrog, M.; Osterle, W.; Sahre, M; Urban, I. Ti/TiN multilayer coatings: deposition technique, characterization and mechanical properties. Surface Coating Technology, v.142/144, p.579-584, 2001.

Herranen, M.; Wiklund, U.; Carlsson, J.O.; Hogmark, S. Corrosion behavior of Ti/TiN multilayer coated tool steel. Surface Coating Technology, v.99, p.191196, 1998.

Wei, C.; Lin, J.F.; Jiang, T.; Ai, C. Tribological characteristics of titanium nitride and titanium carbonitride multilayer films. Thin Solid Films, v.381, p.94-103, 2001.

Andersen, K.N.; Bienk, E.J.; Schweitz, K.O.; Reitz, H.; Chevallier, J.; Kringhoj, P.; Bottiger, J. Deposition, microstructure and mechanical and tribological properties of magnetron sputtered TiN/TiAlN multilayers. Surface Coating Technology, v.123, p.219-226, 2000.

Zhou, Y.; Asaki, R.; Soe, W.; Yamamoto, R.; Chen, R.; Iwabuchi, A. Hardness anomaly, plastic deformation work and fretting wear properties of polycrystalline TiN/CrN multilayers. Wear, v.236, p.159-164, 1999.

Huq, M.Z.; Celis, J.P. Fretting wear of multilayered (Ti, Al)N/TiN coatings in air of different relative humidity. Wear, v.225/229, p.53-64, 1999.

Yashar, P.C.; Sproul, W.D. Nanometer scale multilayered hard coatings. Vacuum, v.55, p.179-190, 1999.

Nordin, M.; Ericson, F. Growth characteristics of multilayered physical vapour deposited TiN/TaNx on high speed steel substrate. Thin Solid Films, v.385, p.174-181, 2001.

Holleck, H.; Schier, V. Multilayer PVD coatings for wear protection. Surface Coating Technology, v.76/77, p.328-336, 1995.

Bull, S.J.; Jones, A.M. Multilayer coatings for improved performance. Surface Coating Technology, v.78, p.173-184, 1996.

Wiklund, U.; Hedenqvist, P.; Hogmark, S. Multilayer cracking resistance in bending. Surface Coating Technology, v.97, p.773-778, 1997.

He, J. L.; Li, W. Z.; Li, H. D.; Liu, C. H. Plastic properties of nano-scale ceramicmetal multilayers Surface Coating Technology, v.103/104, p.276-280, 1998.

Tavares, C.J.; Rebouta, L.; Andritschky, M.; Guimaraes, F.; Cavaleiro, A. Mechanical and surface analysis of Ti0.4Al0.6N/Mo multilayers. Vacuum, v.60, p.339-346, 2001.

Prengel, H.G.; Jindal, P.C.; Wendt, K.H.; Santhanam, A.T.; Hegde, P.L.; Penich, R.M. A new class of high performance PVD coatings for carbidecutting   tools. Surface Coating Technology, v.139, p.25-34, 2001.

Charrier, C.; Jacquot, P.; Denisse, E.; Millet, J. P.; Mazille, H. Aluminum and Ti/Al multilayer PVD coatings for enhanced corrosion resistance. Surface Coating Technology, v.90, p.2934, 1997.

Leoni, M.; Scardi, P.; Rossi, S.; Fedrizzi, L.; Massiani, Y. (Ti,Cr)N and Ti/TiN PVD coatings on 304 stainless steel substrates: texture and residual stress. Thin Solid Films, v.345, p.263-269, 1999.

Nono, M.C.A.; Gomes, G.F.; Vieira, R.A.; Teixeira, J.C.; Algatti, M.A. Material surface modifiers metal by ion implantation and film deposition under development at INPE. In: Brazilian Congress of Materials Science and Engineering, 14., 2001, Águas de Sao Pedro, SP. Proceedings... 2001. 1 CD-ROM. p.44601-44613.

Verardi, P.; Dinescu, M.; Stanciu, C.; Gerardi, C.; Mirenghi, L.; Sandu, V. A parametric study of AlN thin films grown by pulsed laser deposition. Materials Science and Engineering B, v.50, p.223-227, 1997.

Piscanec, S.; Ciacchi, L.C.; Vesselli, E.; Comelli, G.; Sbaizero, O.; Meriani, S.; De Vita, A. Bioactivity of TiN-coated titanium implants. Acta Materialia, v.52, p.1237-1245, 2004.

Huang, H-H.; Hsu, C-H; Pan, S-J; He, J-L; Chen, C-C; Lee, TL. Corrosion and cell adhesion behavior of TiN-coated and ion-nitrided titanium for dental applications. Applied Surface Science, v.244, p.252-256, 2005.

Kola, P.V.; Daniels, S.; Cameron, D.C.; Hashmi, M.S.J. Magnetron sputtering of TiN protective coatings for medical applications. Journal of Materials Processing Technology, v.56, p.422-430, 1996.

Liu, C.; Leyland, A.; Bi, Q.; Matthews, A. Corrosion resistance of multi-layered plasma-assisted physical vapour deposition TiN and CrN coatings. Surface Coating Technology, v.141, p.164-173, 2001.

Andreyev, M.; Anishchik, V. Changes in the structure and properties of TiN coatings irradiated with Ti ions of moderate energies. Vacuum, v.63, p.541-544, 2001.

Peter, S.; Richter, F.; Tabersky, R.; König, U. Optical emission spectroscopy of a PCVD process used for the deposition of TiN on cemented carbides. Thin Solid Films, v.377/378, p.430-435, 2000.

Mogensen, K.S.; Mathiasen C.; Eskildsen, S.S.; Störi, H.; Bottiger, J. The time development of pulsed-DC production plasmas used for deposition of TiN. Surface Coating Technology, v.102, p.35-40, 1998.

Staia, M. H.; Lewis, B.; Cawley, J.; Hudson, T. Chemical vapour deposition of TiN on stainless steel. Surface Coating Technology, v.76-77, p.231-236, 1995.

Kang, T.W.; Kim, T.W. Structural properties of TiN films grown on stainless steel substrates by reactive radio-frequency sputtering technique at low temperature. Applied Surface Science, v.150, p.190-194, 1999.

Guu, Y.Y.; Lin, J.F.; Ai, C. The tribological characteristics of titanium nitride, titanium carbonitride and titanium carbide coatings. Thin Solid Films, v.302, p.193-200, 1997.

Xiao, S.; Lungu, C.P.; Takai, O. Comparison of TiN deposition by rf magnetron sputtering and electron beam sustained arc ion plating. Thin Solid Films, v.334, p.173-177, 1998.

Mori, T.; Fukuda, S.; Takemura, Y. Improvement of mechanical properties of Ti/TiN multilayer film deposited by sputtering. Surface Coating Technology, v.140, p.122-127, 2001.

Ejiri, S.; Sasaki, T.; Hirose, Y. X-ray stress measurement for TiN films evaporated by PVD. Thin Solid Films, v.307, p.178182, 1997.

Lungu, C. P.; Futsuhara, M.; Takai, O.; Braic, M.; Musa, G. Noble gas influence on reactive radio frequency magnetron sputter deposition of TiN films. Vacuum, v.51, n.4, p.635-640, 1998.

Stock, II.R.; Schulz, A.; Kopnarski, M.; Gross, T. Reference materials for composition-depth profiles of TiN, Ti(C,N) and (Ti, Al)N with different chemical composition. Surface Coating Technology, v.98, p.918-924, 1998.

Andrieux, M.; Ducarroir, M.; Beauprez, E. Behavior of radio frequency PACVD bilayers (SiC/TiN) on steel. Thin Solid Films, v.324, p.141-150, 1998.

Yokota, K.; Tamura, S.; Nakamura, K.; Horiguchi, M.; Nakaiwa, H.; Sugimoto, T.; Akamatsu, K.; Nakao, K. Modification of the surfaces of stainless steel during titanium nitride deposition by a dynamic mixing method. Nuclear Instruments and Methods in Physics Research B, v.166/167, p.87-91, 2000.

Bergmann, E.; Van Der Kolk, G.I.; Buil, B.; Hurkmans, T. The next generation of deposition equipment for wear protection coatings. Surf. Coat. Technol., v.114, p. 101-107, 1999.

Sarwar, M. Applications of advanced surface engineering treatments to multipoint cutting edges. Surface Coating Technology, v.108/109, p. 612-619, 1999.

Mattox, D. M. The application of plasmas to thin film deposition processes. Plasma-Surface Interactions and Processing of Materials, v.176, p.377-399, 1990.

Friedrich, C.; Berg, G.; Senf, J.; Broszeit, E. Reliable PVD coatings on components: aspects of depositionand characterization for quality management. Surface Coating Technology, v.112, p. 152-161, 1999.

Jehn, H.A. PVD and ECD-competition, alternative or combination? Surface Coating Technology, v.112, p. 210-216, 1999.

Heidsieck, H. Status of vacuum and plasma technology. Surface Coating Technology, v.112, p. 324-338, 1999.

Perry, A. Process and property relationships in hard coatings made by plasma-and ion-assisted methods. Surface Coating Technology, v.52, p. 261-267, 1992.

Shohet, J. L. Plasma-aided manufacturing. IEEE Trans. on Plasma Science, v.19, n.5, p.725-733, 1991.

Mattox. D. M. Plasma-surface interactions and processing of materials. Applied. Sciences Serie

E, v.176, p.377-399, 1990.

Brühl, S. P. PVD plasma coatings. Buenos Aires: CNEA, 1998. p. 71-86.

Pascual, R. Propriedades y caracterización de recubrimientos. Buenos Aires: CNEA, 1998. p.139-156.

Sproul, W. Physical vapor deposition tool coating. Surface Coating Technology, v.81, p.1-7, 1996.

Eskildsen, S.S.; Mathiasen, C.; Foss, M. Plasma CVD: process capabilities and economic aspects. Surface Coating Technology, v.116/119, p.18-24, 1999.

Verhoeven, J. D. Scanning electron microscopy. In: Wang, R.E. (ed). Metals handbook: materials characterization. 9.ed. Ohio: American Society for Metal, 1986. p.491-515.

Tufts, C.F. Transmission electron microscopy and electron diffraction. In: Hench, L.L.; Gould, R.W. (eds).

Characterization of ceramics. [S.l.]: [S.n.], 1971. p.177-217.

Heinrich, K.F.J., Newbury, D. E. Electron probe X-ray microanalysis. In: Wang, R.E. (ed). Metals handbook: materials characterization. 9.ed. Ohio: American Society for Metal, 1986. p. 516-545.

Goehner, R. P., X-ray powder diffraction. In: Wang, R.E. (ed). Metals handbook: materials characterization. 9.ed. Ohio: American Society for Metal, 1986. p.333343.

Gould, R. W. X-ray diffraction. In: Hench, L.L.; Gould, R.W. (eds). Characterization of ceramics. [S.l.]: [S.n.], 1971. p.135175.

Klug, H.; Alexander, L. X-ray diffraction procedures. New York: John Willey & Sons, 1954. p.491-538.

Rankin, P. Chemical analysis. In: Hench, L.L.; Gould, R.W. (eds). Characterization of ceramics. [S.l.]: [S.n.], 1971. p.9-36.

Gould, R. W. X-ray spectroscopy. In: Hench, L.L.; Gould, R.W. (eds). Characterization of ceramics. [S.l.]: [S.n.], 1971. p.39-87.

Bates, S. R. Scanning electron microscopy. In: Hench, L.L.; Gould, R.W. (eds). Characterization of ceramics. [S.l.]: [S.n.], 1971. p.419-434.

Sarver, L. W. SEM and EDS analyze materials. Advanced Materials & Precesses, v.149, p.19-21, 1996.

Kelly, T. F. The scanning electron microscope. MME362- Physical Metallurgy Laboratory. University of Wisconsin- Madison. Department of Metallurgical and Mineral Engineering, p.2-9 (Lecture notes).

Metzner, H.; Hahn, Th.; Gossla, M.; Conrad, J.; Bremer, J. - H. Rutherford backscattering spectroscopy of rough films: Experimental aspects. Nuclear Instruments and Methods in Physics Research B, v.134, p.249-261, 1998.

Metzner, H.; Gossla, M.; Hahn, Th. Rutherford backscattering spectroscopy of rough films: Theoretical considerations. Nuclear Instruments and Methods in Physics Research B, v.124, p.567-574, 1997.

JEOL JSM5310 Scanning Electron Microscope (SEM) manual.

Ewing, G.E. Instrumental methods of chemical analysis. Sao Paulo: Edgard Blucher, 1972. v.1.

Ipohorski, M. Técnicas de microscopia electronica. Buenos Aires: CNEA, 1998. p. 179-213.

Sedin, D.L.; Rowlen, K.L. Influence of tip size on AFM roughness measurements. Applied Surface Science, v.182, p.40-48, 2001.

Poon, C.Y.; Brushan, B. Comparison of surface roughness measurements by stylus profiler, AFM and non-contact profiler. Wear, v.190, p.76-88, 1995.

Chastain, J. Handbook of X-ray photoelectron spectroscopy. [S.l.]: PerkinElmer, 1992.

Smith, G.C. Surface analysis by electron spectroscopy. New York: Plenum, 1994.

Strafford, K.N.; Subramanian, C. Surface engineering : an enabling technology for manufacturing industry. Journal of Material Processing Technology, v.53, p.393-403, 1995.

Binh, V. T. Surface mobilies on solid materials: fundamental concepts and applications. [S.l.]: Plenum Press, 1983.

Mclean, M. Thermodynamics of interfaces. In: (ed).

The Science of Interfaces in Materials Engineering, [S.l.]: [S.n.], 1988. p.8-27.

Mclean, M. Interface in materials. In: (ed). The

science of interfaces in materials engineering. Belgium: Proceedings of The Colloquium Brusels, 1988. p.7-27.

Tabor, D. Interaction between surfaces : adhesion and friction. In: Blakely, J.M. (ed.). Surface physics of materials. New York: Academic Press, 1975. p.476529.

Spaepen, F. Interfaces and stresses in thin films. Acta Materialia, v.48, p.31-42, 2000.

Mattox, D.M. Thin film adhesion and adhesive failure - A perspective. In: Mittal, K.L. (ed). Adhesion measurement of thin films, thick films and bulk coatings. [S.l.]: American Society for Testing and Materials, 1978. p. 54.

Alvarez, F. J. CVD plasma coating techniques. Buenos Aires: CNEA, 1998. p.47-69.

Hill, R. J. Electron beam technology. In: (ed). Physical vapor deposition, temescal. Berkeley: BOC Group, 1986. p. 17-105.

Engers, B.; Bauer, H. U. Cost-effective PVD coatings in Batch systems. Surface Coating Technology, v.116/119, p.705-710, 1999.

Hill, R. J. Sputtering systems. In: (ed). Physical vapor deposition, temescal. 2.ed. Berkeley: BOC Group, 1986. p.111-189.

North, B. Six issues for the hard coatings community. Surface Coating Technology, v.106, p. 129-134, 1998.

Pierson, H. O. Handbook of chemical vapor deposition (CVD): principles, technology and applications. [S.l.]: Noyes, 1992.

TTC-Thermal treatments-Advanced coating, PVD-Physical vapor deposition. Available at: <http://www.ttc-spain.com>. Accessed on: March 25, 2004.

Kern, W.; Schuegraf, K.K. Deposition technologies and applications: introduction and overview. In: Schuegraf, K.K. (ed). Handbook of thin-film deposition processes and techniques. [S.l.]: [S.n.], 1988. p.1-25.

Ohring, M. The materials science of thin films. New York: Academic Press, 1992.

Seah, M.P.; Dench, W.A. Quantitative electron spectroscopy of surfaces: a standard data base for electron inelastic mean free paths in solids. Surface Interface Analysis, v.1, p.2-11, 1979.

Harrison, K.; Hazell, L.B. The determination of uncertainties in quantitative XPS AES and its impact on data acquisition strategy. Surface Interface Analysis, v.18, p.368-378, 1992.

Piorino, F.N. On the variables that affect the evaluation of the modulus of rupture in bending tests. 1990. Dissertation (Master's Degree in Materials Engineering) - Federal University of Sao Carlos, Sao Carlos, 1990.

Folsom, C.A.; Zok, F.W.; Lange, F.F. Flexural properties of brittle multilayer materials: II, experiments. Journal of American Ceramic Society, v.77, n.8, p.2081-2087, 1994.

The American Society for Testing and Materials (ASTM). Standard test methods for bend testing of metallic apartment materials for spring applications involving static loading. Philadelphia: ASTM, 1990.

Beer, F.P.; Johnston, E.R. Resistance of materials. 2.ed. Sao Paulo: Mac Grow-Hill, 1989.

Heidsieck, H. Status of vacuum and plasma technology. Surface Coating Technology, v.112, p. 324-338, 1999.

Uddehol Apos Especiais. Manual de apos. [S.l.]: Uddeholm, 2000.

Doolittle. Computer Graphic Service. 1995. Available at: <http://www.genplot.com>. Accessed on: March 26, 2004.

Uglov, V.V.; Fedotova, J.A.; Kuleshov, A.K.; Danilyuk, A.L.; Kvasov, N.T.; Günzel, R.; Reuther, R.; Richter, E. Evolution of microstructure of instrumental AISI M2 steel after plasma immersion nitrogen and carbon implantation. Surface Coating Technology, v.136, p.226-230, 2001.

Maissel, L.; Gland, R. Handbook of thin film technology. New York: McGrawHill, 1970.

Li, H.; Selvaduray, G. Ellimgham diagram web Project. Available at   : . Accessed on March 25, 2004.

Kubaschewski, O.; Alcock, C. Metallugical thermochemistry. 5.ed. London: Pergamon, 1989.

Vieira, R.A.; Nono, M.C.A. Study of film defects formed during the deformation of titanium coating on 304 SS substrates. Acta Microscópica, v. 9, p. 285-286, 2001.

Printed by Books on Demand GmbH, Norderstedt / Germany